D1327592

Critical effects in semiclassical scattering, in which the standard approximations break down, are associated with forward peaking, rainbows, glories, orbiting and resonances. Besides giving rise to beautiful optical effects in the atmosphere, critical effects have important applications in many areas of physics. However, their interpretation and accurate treatment is difficult. This book, based on the Elliott Montroll Lectures, given at the University of Rochester, deals with the theory of these critical effects.

After a preliminary chapter in which the problem of critical effects is posed, the next three chapters on coronae, rainbows and glories are written so as to be accessible to a broader audience. The main part of the book then describes the results obtained from the application of complex angular momentum techniques to scattering by homogeneous spheres. These techniques lead to practically usable asymptotic approximations, and to new physical insights into critical effects. A new conceptual picture of diffraction, regarded as a tunneling effect, emerges. The final two chapters contain brief descriptions of applications to a broad range of fields, including linear and nonlinear optics, radiative transfer, astronomy, acoustics, seismology and atomic, nuclear and particle physics. The book intends to convey the basic concepts and physical interpretations that emerge from the new approach, rather than the complete formalism. Readers are referred to the literature for more detailed results.

This volume will be of use to graduate students and researchers in many areas of physics. Nonspecialists interested in the explanation of the beautiful optical phenomena which occur in the atmosphere may also profit from the introductory chapters.

Elliott W. Montroll, whose contributions to science are honored in this lecture Series in Mathematical Physics, was from 1966 until his retirement in 1981 the Einstein Professor of Physics at the University of Rochester, where the Montroll Lectures are held. Montroll was also at various times Distinguished Professor in the Institute for Physical Science and Technology of the University of Maryland, Distinguished Professor of Physics of the University of California at Irvine, Director of General Sciences at the IBM Corporation, and Vice President for Research of the Institute for Defense Analyses. At Rochester he was founder of the Institute of Fundamental Studies and maintained a vigorous and varied research program.

Montroll's research career began in 1941 with the earliest publication of the diagrammatic resummation procedure that subsequently became so prevalent in theoretical physics, and the introduction of the moment trace method for evaluating the vibrational frequency spectrum of a lattice. Among his other contributions to mathematical physics were his development of the transfer matrix method for calculating the partition function of an interacting lattice system; the exact calculation of the spin–spin correlation functions of the two-dimensional Ising lattice using the Pfaffian method; the development during the Second World War of the control system used to stabilize the cascade separation of uranium isotopes; the first exact evaluation of the vibrational frequency spectrum of a two-dimensional lattice and pioneering work on the theory of vibrations of crystals with defects; major contributions to random walk theory and its application to physical and chemical problems, for example, the exact solution to Polya's problem about the probability of return to the origin of a three-dimensional lattice. His introduction with George Weiss of the notion of a continuous-time random walk and the associated pausing time distribution led to ground-breaking advances in the theory of transport and relaxation in disordered systems.

Among his many other activities, Montroll was twice the Lorentz Professor at Leiden, the Gibbs Lecturer for the American Mathematical Society, founder and first editor of the Journal of Mathematical Physics, recipient with Robert Herman of the Lancaster Prize and a member of the US National Academy of Sciences. His influence on the development of mathematical and chemical physics was widely appreciated through his exceptionally lucid articles and lectures, blending common sense and beautiful mathematics.

The Montroll Memorial Lecture Series has been established by Elliott Montroll's friends to provide a forum for the presentation of new developments and coherent overviews in mathematical physics. The lectures are given at the University of Rochester. It is appropriate that the Lectures will be available in book form as a continuing representation of Montroll's vitality and curiosity and intellectual commitment to the understanding and explaining of the world of science.

A. Das, J. H. Eberly, M. M. Shahin, H. M. Van Horn

MONTROLL MEMORIAL LECTURE SERIES IN
MATHEMATICAL PHYSICS: 1

*Diffraction Effects in Semiclassical Scattering*

# Diffraction Effects in Semiclassical Scattering

H. M. NUSSENZVEIG

*Department of Physics*
*Pontifical Catholic University of Rio de Janeiro*

CAMBRIDGE
UNIVERSITY PRESS

Published by the Press Syndicate of the University of Cambridge
The Pitt Building, Trumpington Street, Cambridge CB2 1RP
40 West 20th Street, New York, NY 10011-4211, USA
10 Stamford Road, Oakleigh, Victoria 3166, Australia

First published 1992

Printed in Great Britain at the University Press, Cambridge

*A catalogue record of this book is available from the British Library*

*Library of Congress cataloguing in publication data*
Nussenzveig, H. M. (Herch Moysés)
    Diffraction effects in semiclassical scattering / H. M.
    Nussenzveig.
        p.    cm.
    Includes bibliographical references and index.
    ISBN 0 521 38318 8
    1. Scattering (Physics).    2. Diffraction.    3. Angular momentum.
    4. Mathematical physics.    I. Title.
    QC20.7.S3N87 1992
    530.1′5–dc20    91-30177 CIP

ISBN 0 521 38318 8 hardback

# Contents

# Preface

*The full solution of the problem presented by spherical drops of water*
*would include the theory of the rainbow, and if practicable at all would be*
*a very complicated matter.*

(Rayleigh 1899)

The standard semiclassical approximations in scattering theory break down in four well-known situations, that correspond to some of the most interesting scattering effects. In order to treat them, one must be able to handle diffraction, a notoriously difficult subject involving the dynamical aspects of wave propagation.

The basic theme of this book is how to deal with these critical effects. Fortunately for the theorist, there exists an exactly soluble model that exhibits all of them, the scattering of light by a homogeneous spherical particle. Some of the most beautiful phenomena in nature, coronae, rainbows and glories, are visual manifestations of the critical effects contained within this model. The model is not only soluble: it describes with extremely high accuracy the actual behavior of small liquid droplets.

The formally exact solution of the light scattering problem, obtained by Mie at the beginning of this century (and even earlier by Lorenz), is in the form of an infinite series, the well-known partial-wave series; analogous problems for elastic waves had been solved in this form several decades before. During this period spanning over a century, many important contributions to mathematical physics, particularly in connection with special functions and asymptotic approximation methods, originated in the treatment of these problems.

Unfortunately, the series converges very slowly under semiclassical conditions. In view of the manifold practical applications, substantial effort has been devoted to developing fast computer programs for the numerical summation of the series solution. Notwithstanding the widespread use of such programs, this approach leads to both practical and theoretical difficulties.

In practice, the results display very rapid and complicated fluctuations that depend sensitively on all parameters. One usually wants to get rid of these fluctuations, retaining only contributions that survive after some statistical averaging over the parameters. On the theoretical side, the physical content of the formal solution remains buried within the series representation. The challenge is to retrieve it. Similar difficulties are found in semiclassical problems that arise in many other fields.

My own involvement with such problems began more than two decades ago, prompted by a desire to understand the physical interpretation of Regge poles. Complex angular momentum theory arose from attempts by Poincaré and Watson, at the beginning of this century, to overcome the slow convergence problem of the partial-wave series. It is a very powerful tool, fully capable of fulfilling this task.

The motivation for writing this book was provided by the Montroll Lectures that I gave at the University of Rochester in 1988. The lecturer is required, in addition to specialized talks on his chosen topic, to address one lecture to a general audience. Mine was entitled 'The Universe of Light in a Droplet', a title borrowed partly from Bragg's beautiful lectures at the Royal Institution and partly from the idea that the elucidation of the phenomena resulting from the interaction between light and a water droplet has required the utilization of all successive theories about the nature of light, from early corpuscular theories to present-day quantum theory. A considerable amount of material, including new developments, has been added to the original set of lectures.

In planning the book, I decided to retain the format of the Montroll Lectures in at least one respect: after a preliminary chapter in which the problem of critical effects is posed, the next three chapters, on coronae, rainbows and glories, are written so as to be accessible to a broader audience. They can be read independently, but they are also meant as an introduction to and (together with Chapter 5) an overview of the phenomena to be treated.

The main part of the book (Chapters 6 through 14) describes the results obtained by the application of complex angular momentum techniques to scattering by homogeneous spheres. They lead to accurate and practically usable asymptotic approximations, as well as to new physical insights about the critical effects. A new conceptual picture of diffraction, regarded as a tunneling effect, emerges.

The mathematical developments and derivations cannot be presented in a compact monograph; the reader is referred to the literature for more detailed results. What the book intends to convey is the physical picture that emerges from the new approach: basic concepts and ideas, rather than formalism. An informal style is employed, as well as a variety of illustrations.

The last two chapters contain brief descriptions of applications to a broad range of fields: linear and nonlinear optics, radiative transfer, astronomy, acoustics, seismology, atomic, nuclear and particle physics.

It is hoped that the book may be useful to research workers in these areas, as well as to mathematical physicists. The material is accessible to graduate students; non-specialists interested in the explanation of the beautiful optical phenomena in the atmosphere may profit from the introductory chapters.

In dedicating the book to the memory of Elliott Montroll, who included contributions to this subject among his manifold interests, and who was a great admirer of Lord Rayleigh, one of the grandfathers of scattering by spheres, I keep a vivid recollection of how thrilled he was when I told him about quasichaotic features of the ripple fluctuations in Mie scattering. He believed that such features would eventually emerge from the study of any problem, if one probed deep enough.

I have been very fortunate in dealing with this theme to count on the invaluable collaboration of Vijay Khare, Warren Wiscombe, Nelson Fiedler-Ferrari and Luiz Gallisa Guimarães. Warren Wiscombe, together with Dennis Chesters, also patiently dealt with my computer illiteracy so as to enable me to compose this book, including most of the illustrations, entirely on a personal computer. I thank my wife, Micheline, for her support throughout the years of work summed up in the present monograph and for her patience and understanding during its preparation.

An early version of these Montroll Lectures was given at the Collège de France in 1985. I would like to thank the Montroll Lectures Committee for inviting me to deliver the lectures on which this book is based, and to express my thanks to Albert Arking for his hospitality at NASA Goddard Space Flight Center, where most of the book was written, and to Claude Cohen-Tannoudji and Serge Haroche for their hospitality at the Ecole Normale Supérieure, where it was finished. Support by the US National Research Council, by the French CNRS and by the Brazilian agencies CNPq, CAPES and FINEP is gratefully acknowledged.

H. M. Nussenzveig
Paris

# Critical effects in semiclassical scattering

*Ordinary mechanics : Wave mechanics*
*= Geometrical optics : Undulatory optics.*
(Schrödinger 1928)

In this chapter, the basic critical scattering effects to be dealt with in this book are introduced. They appear first as singularities, at the level of classical mechanics or geometrical optics. The transition from semiclassical to classical scattering is then outlined, and the need for more refined approximations is brought out.

## 1.1 Classical scattering

In classical mechanics, the scattering of a nonrelativistic particle of mass $m$ by a central potential $V(r)$ is determined by the conservation laws of angular momentum $L$ and energy $E$,

$$L = mr^2 \dot{\phi}, \quad E = \tfrac{1}{2} m \dot{r}^2 + U_L(r) \tag{1.1}$$

where $\phi$ is the polar angle in the scattering plane and $U_L(r)$ is the *effective potential*,

$$U_L(r) = V(r) + L^2 / (2mr^2) \tag{1.2}$$

in which the last term is the contribution from the 'centrifugal potential'.

From (1.1), we can solve for $\dot{\phi}/\dot{r} = \mathrm{d}\phi/\mathrm{d}r$. The radial motion in scattering ranges from $r = \infty$ to the largest root $r_0$ of $\dot{r} = 0$, which defines the *classical distance of closest approach* (outermost radial turning point, see fig. 1.1), and then back from $r_0$ to $\infty$. Integrating $\mathrm{d}\phi/\mathrm{d}r$ between these extreme values, with the initial condition $\phi = \pi$ at $r = \infty$, we find the overall *classical deflection angle* $\Theta$:

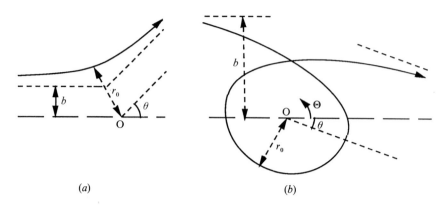

Fig. 1.1. *(a)* Repulsive interaction; *(b)* Attractive interaction. $b$ = impact parameter; $r_0$ = classical distance of closest approach. In *(a)*, $\Theta = \theta$; in *(b)*, $\Theta = -2\pi - \theta$.

$$\Theta(L) = \pi - 2L \int_{r_0}^{\infty} \frac{dr}{r^2 \left\{ 2m\left[ E - U_L(r) \right] \right\}^{\frac{1}{2}}} \tag{1.3}$$

which is a function of $L$, or, equivalently, of the *impact parameter* $b = L/p$ associated with the collision (fig. 1.1), where $p$ is the magnitude of the incident linear momentum.

As illustrated in fig. 1.1, $\Theta$ ranges from 0 to $\pi$ for a repulsive interaction, but it can take arbitrarily large negative values for an attractive one, since the particle may then go around the scattering center any number of times before its final emergence. Thus, $\Theta$ is related to the *scattering angle* $\theta$ by

$$\Theta + 2n\pi = \pm\theta \qquad (n = 0, 1, 2, \ \ldots) \tag{1.4}$$

where $n$ is a nonnegative integer such that $\theta$ falls within its physical range of variation, $0 \le \theta \le \pi$.

The cross-sectional area associated with particles with impact parameters between $b$ and $b + db$ is $2\pi b|db|$. Thus, if they are scattered within a solid angle $d\Omega = 2\pi \sin\theta\, d\theta$ the corresponding contribution to the differential scattering cross section is

$$\frac{d\sigma}{d\Omega} = \frac{b}{\sin\theta} \left| \frac{db}{d\theta} \right|$$

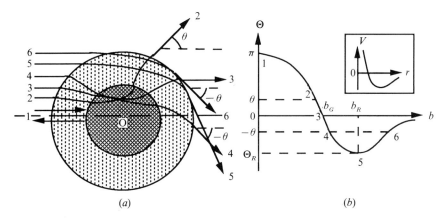

(a)                                                    (b)

Fig. 1.2. *(a)* Sketch of various paths for an attractive potential ⬚ with a repulsive core
▨. *(b)* Corresponding deflection function $\Theta(b)$. The inset shows the typical shape
assumed for the potential. Path 3 is a glory path and path 5 is a rainbow path. The
repulsive path 2 and the two attractive paths 4 and 6 lead to the same scattering angle $\theta$.

As will be seen shortly, there may in general exist several different paths
leading to the same scattering angle. It follows that the *classical
differential cross section* in the direction $\theta$ is given by

$$\frac{d\sigma}{d\Omega}(\theta) = \sum_j \frac{b_j(\theta)}{\sin\theta} \left| \frac{d\theta}{db_j} \right|^{-1} \qquad (1.5)$$

summed over all impact parameters $b_j$ that lead to the same scattering
angle $\theta$. The derivative that appears in (1.5) is equivalent to the Jacobian
of the transformation relating the variables $\theta$ and $b$.

Fig. 1.2*(a)* illustrates a variety of paths for a type of potential
commonly found in atomic and nuclear physics, an attractive interaction
with a repulsive central core; fig. 1.2*(b)* shows the associated *deflection
function* $\Theta(b)$. Thus, for a central collision (path 1), the particle is turned
back by the repulsive core: $\Theta = \pi$ for $b = 0$. As $b$ increases, the repulsive
interaction at first prevails, so that $\Theta > 0$ (path 2), but it can eventually be
compensated by the effect of the attractive one, so that the particle
emerges undeflected (path 3). For larger impact parameters, the overall
effect is attractive ($\Theta < 0$), but the attraction usually decreases at larger
distances, so that, after going through a largest negative deflection (path
5), $\Theta(b)$ increases, tending to zero as $b \to \infty$. We see in fig. 1.2 that there
are three different paths leading to the same scattering angle $\theta$ for the

value of $\theta$ shown: the repulsive path 2 and the attractive paths 4 and 6. Their respective contributions must be summed in (1.5).

From (1.5) we can already infer the existence of three different *singularities in classical scattering*, associated with directions in which the differential cross section diverges:

(I) *Rainbow scattering*. This takes place when the deflection function goes through a maximum or a minimum, as it does at $\Theta_R$ in fig. 1.2(*b*). The divergence occurs at the *rainbow angle* $\theta_R$, where

$$\left(d\theta/db\right)_{\theta=\theta_R} = 0 \qquad (1.6)$$

Entering a parabolic approximation to the deflection function around $\theta_R$ into (1.5), it is readily found that $d\sigma/d\Omega$ diverges like $|\theta - \theta_R|^{-1/2}$ as $\theta$ approaches $\theta_R$ from the side that is accessible by classical paths (these paths 'turn back' after the extremal deflection, so that they do not contribute on the other side). The denomination 'rainbow scattering' originates from the analogy with the optical rainbow (cf. Chapter 3).

(II) *Glory scattering*. If the deflection function goes through zero or a negative multiple of $\pi$ for a nonzero value of the impact parameter $b_G$ (cf. fig. 1.2),

$$\Theta(b_G) = n\pi \ ( \ n = 0, -1, -2, \ ...; \ b_G \neq 0 \ ) \qquad (1.7)$$

the differential cross section, by (1.5), diverges like $(\sin \theta)^{-1}$ either in the exactly forward direction (*n* even) or in the exactly backward direction (*n* odd). This is known as forward or backward glory scattering, respectively, although the optical phenomenon of the meteorological glory, described in Chapter 4, has an entirely different origin. Note that, by (1.3), $\Theta = \pi$ may occur only for $b = 0$, as in fig. 1.2, so that it does not lead to glory scattering.

(III) *Forward peaking*. If the scattering potential has a tail extending to infinity (e.g., decaying exponentially or like some inverse power of the distance), particles with arbitrarily large impact parameters will still undergo small deflections, so that one would expect a clustering of such paths about $\theta = 0$, leading to a divergent forward differential cross section. This indeed follows from (1.5), for the contribution from paths such that $b(\theta) \to \infty$ and $\theta \to 0$, while $|d\theta/db|$ remains bounded. For a cutoff potential, such as a hard core, this argument no longer holds, but

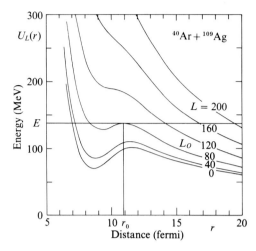

Fig. 1.3. Family of effective nucleus–nucleus potentials $U_L(r)$ for $^{40}$Ar + $^{109}$Ag (after Myers 1974). For $L = L_O$, orbiting takes place at the energy $E$.

one would still expect an anomaly in forward scattering, as will be seen later.

Besides these singularities, that follow from (1.5), the possibility of a fourth anomalous situation can be inferred from (1.3). To see how it arises, we consider a family of effective potentials $U_L(r)$ for a range of values of the angular momentum $L$, such as that illustrated in fig. 1.3, that arises from a potential $V(r) = U_0(r)$ employed in nuclear physics, of a type similar to that considered in fig. 1.2.

(IV) *Orbiting.* For a given $L_O$ in the family such that $U_{L_O}(r)$ goes through a maximum at $r = r_O$, consider a path with energy $E$ such that

$$U_{L_O}(r_o) = E, \quad \left(dU_{L_O}/dr\right)_{r=r_o} = 0 \tag{1.8}$$

so that the energy sits right at the top of the barrier (fig. 1.3). This amounts to the existence of an unstable circular orbit with radius $r_O$.

Under these conditions, the integral in (1.3) diverges at its lower limit [if the peak in the potential is parabolic, it can be verified that the divergence is logarithmic (Ford & Wheeler 1959)]. The result $\Theta(L_O) \rightarrow -\infty$ for the deflection angle means that an incident particle with this energy and angular momentum will spiral indefinitely around the scattering center, taking an infinitely long time to reach the top of the barrier; hence the name 'orbiting'. The inverse function $b(\theta)$ becomes infinitely many-valued, so that there are an infinite number of branches that contribute to the differential cross section (1.5).

Note also that the existence of a pocket in the effective potential for $r < r_O$ implies that a particle with energy below the barrier top can be captured into an oscillatory orbit within the potential, so that the unstable circular orbit acts like a separatrix between bounded and unbounded paths.

## 1.2    Semiclassical scattering

Under what conditions are classical concepts and results relevant in quantum scattering? Classical mechanics is expected to emerge as a limiting case of quantum mechanics in the (formal) limit as $\hbar \rightarrow 0$. However, this is an extremely complicated limit, both mathematically and from a physical standpoint.

There is a very close analogy between this situation and that found in optics, when one considers geometrical optics as a limiting case of wave optics. This is no accident: Hamiltonian mechanics was built upon the analogy with geometrical optics (Whittaker 1927, Goldstein 1957), and Schrödinger's wave mechanics was developed with the help of the analogy with wave optics (Sommerfeld 1954, Born & Wolf 1959).

In Hamilton's analogy, the path of a nonrelativistic particle with energy $E$ in a central potential $V(r)$ is placed into correspondence with a geometrical-optic light ray in an inhomogeneous medium with the refractive index

$$N = \sqrt{1 - [V(r)/E]} \qquad (1.9)$$

and the principle of least action corresponds to Fermat's principle (the action is the analogue of the optical path).

Similarly, in Schrödinger's extension, the analogue of the optical reduced wavelength is the de Broglie reduced wavelength

$$\lambda \equiv 1/k = \hbar/p = \hbar \big/ \{2m[E - V(r)]\}^{\frac{1}{2}} \qquad (1.10)$$

and the analogue of the circular frequency is given by Einstein's relation $\omega = E/\hbar$, so that, by (1.9), one is dealing with a dispersive medium. From this point of view, the basic new effects that come into play in quantum mechanics correspond to interference and diffraction.

In a semiclassical description of scattering, we still want to be able to speak of particle paths and their individual contributions. However, because they are now contributions to the scattering amplitude for de Broglie waves, they will interfere, so that, instead of a sum of incoherent contributions to the differential cross section like (1.5), one will get a coherent superposition of amplitudes. Thus, when more than one path contributes, there will generally arise interference oscillations. Such oscillations, however, usually (though not always) take place on a very fine scale, below the threshold of resolution of ordinary detectors, so that the interference is washed out by the detection process, eventually leading back to the incoherent result (1.5). It is by such a subtle process that the classical limit is attained, just as one does not ordinarily detect wave-optical effects, leading to a geometrical-optic description of light propagation.

One basic requirement for the applicability of geometrical optics (classical mechanics) is that the short-wavelength limit must be approached: the wavelength must be much smaller than all relevant dimensions. Some of the necessary conditions for the validity of semiclassical approximations to quantum scattering can be obtained by examining under what circumstances one can speak of a well-defined path and a well-defined deflection (Williams 1945).

We assume that the interaction is characterized by a range $a$. To have a well-defined path for an incident particle requires that the uncertainty $\delta b$ in the impact parameter be at most of order $a$. According to the uncertainty relation, this implies an uncertainty $\delta p_\perp$ in the transverse momentum at least of order $\hbar/a$, leading to an uncertainty in scattering angle $\delta\theta \sim \delta p_\perp/p \gtrsim \hbar/pa = 1/ka$, where $k = 1/\lambda$ is the wave number. For the deflection to be well defined, we need to have $\delta\theta \ll 1$, $\delta\theta \ll \theta$, so that we get the conditions

$$\beta \gg 1 \qquad\qquad (1.11)$$

and

$$\theta \gg 1/\beta \qquad\qquad (1.12)$$

where we have introduced a dimensionless parameter that plays a fundamental role, the *size parameter*

$$\beta \equiv ka \qquad (1.13)$$

a measure of the range of the interaction in wavelength units.

Condition (1.11) agrees with the association between semiclassical approximations and the short-wavelength limit. Condition (1.12) excludes a neighborhood of the forward direction which is related with the forward peaking effect already found in Section 1.1.

The differential cross section is related to the quantum scattering amplitude $f(k,\theta)$ by $d\sigma/d\Omega = |f(k,\theta)|^2$, and $f(k,\theta)$ is given by the well-known *partial-wave expansion* (Messiah 1959, Mott & Massey 1965, Schiff 1968)

$$f(k,\theta) = (ik)^{-1} \sum_{l=0}^{\infty} (l+\tfrac{1}{2})[S_l(k)-1]P_l(\cos\theta) \qquad (1.14)$$

where $P_l$ is the Legendre polynomial and

$$S_l(k) = \exp[2i\eta_l(k)] \qquad (1.15)$$

is the $S$-function for the $l$th partial wave, associated with the phase shift $\eta_l(k)$. The angular momentum associated with the $l$th partial wave is

$$L = \sqrt{l(l+1)}\hbar \sim (l+\tfrac{1}{2})\hbar \qquad (1.16)$$

where the last approximation holds for large $l$.

Interpreting an incident plane wave in terms of a particle beam, the $l$th partial wave corresponds to impact parameters concentrated mainly within a circular ring with inner and outer radii $l\lambdabar$ and $(l+1)\lambdabar$, respectively (Blatt & Weisskopf 1952). In a semiclassical approximation, the *localization principle* (van de Hulst 1957) associates with the $l$th partial wave an impact parameter $b_l$ such that $L = pb_l$, where $L$ is given by the last member of (1.16), so that $b_l$ represents the midradius of the above circular ring, namely,

$$b_l = (l+\tfrac{1}{2})/k \qquad (1.17)$$

We expect that incoming paths will interact significantly with the scatterer, leading to appreciable phase shifts, when $b_l \lesssim a$, i.e., by (1.17), when $l + \tfrac{1}{2} \lesssim ka$. Thus, *the typical number of partial waves that must be retained in the partial-wave expansion (1.14) under semiclassical conditions is of the order of the size parameter* $\beta$, *i.e., it is very large.* This poor numerical convergence of the partial-wave expansion, and more generally of other corresponding eigenfunction expansions not associated with scattering, is characteristic of semiclassical situations, and it is the main difficulty that one has to contend with in trying to derive accurate semiclassical approximations.

It is useful to give a brief outline of the main steps involved in the transition from the partial-wave expansion to the classical differential cross section (1.5) (Ford & Wheeler 1959; for a more complete discussion, see Berry 1969 and Berry & Mount 1972). Assuming that the potential $V(r)$ is slowly-varying within a de Broglie wavelength (short-wavelength limit), one replaces the phase shift in (1.15) by its *WKB (Wentzel, Kramers and Brillouin) approximation*, in the form obtained by Langer (1937),

$$\hbar \eta_l^{\text{WKB}} = \int_{r_0}^{\infty} \left[ p_l(r) - p \right] dr - pr_0 + L\frac{\pi}{2} \tag{1.18}$$

where

$$p = \sqrt{2mE}, \quad L = \left(l + \tfrac{1}{2}\right)\hbar \tag{1.19}$$

and

$$p_l(r) = \left\{ 2m\left[ E - U_L(r) \right] \right\}^{\frac{1}{2}} \tag{1.20}$$

with $U_L(r)$ given by (1.2), is the *effective radial momentum*. In (1.18), $r_0$ is the outermost radial turning point of the potential, as in (1.3); note also that $L$ in (1.19) corresponds to the last member of (1.16).

Since most of the partial wave contributions arise from large values of $l$, one can also employ the asymptotic approximation to $P_l(\cos\theta)$ valid for $l\sin\theta \gg 1$ (Abramowitz & Stegun 1965)

$$P_l(\cos\theta) \approx \left[ \frac{2}{\pi\left(l + \tfrac{1}{2}\right)\sin\theta} \right]^{\frac{1}{2}} \cos\left[ \left(l + \tfrac{1}{2}\right)\theta - \frac{\pi}{4} \right] \tag{1.21}$$

provided that one excludes a neighborhood of the forward [cf. (1.12)] and backward directions. By the completeness relation for Legendre polynomials, we have

$$\sum_{l=0}^{\infty} \left(l + \tfrac{1}{2}\right)P_l(1)P_l(x) = \sum_{l=0}^{\infty} \left(l + \tfrac{1}{2}\right)P_l(x) = \delta(1 - x) \tag{1.22}$$

where $\delta$ denotes Dirac's delta function, so that the exclusion of near-forward directions allows us to neglect the term '1' within the square brackets in the partial-wave expansion (1.14).

The roughest approximation to the partial-wave sum when many partial waves contribute is obtained through the replacement of the sum by an integral. With the above substitutions, (1.14) becomes

$$f(k,\theta) \approx (ik)^{-1}(2\pi\sin\theta)^{-\frac{1}{2}}$$
$$\times \int_0^\infty d\lambda\,\sqrt{\lambda}\,[\exp(i\phi_+) + \exp(i\phi_-)] \qquad (1.23)$$

where

$$\lambda \equiv l + \tfrac{1}{2} \qquad (1.24)$$

now becomes a continuous variable, and

$$\phi_\pm(\lambda,\theta) = 2\eta_\lambda^{\text{WKB}} \pm \left(\lambda\theta - \frac{\pi}{4}\right) \qquad (1.25)$$

where $\eta_\lambda^{\text{WKB}}$ is the interpolation for continuous angular momentum of the WKB phase shift, obtained from (1.18) by the substitution (1.24). We refer to $\lambda$ as 'angular momentum' (measured in units of $\hbar$) for simplicity.

The integrand of (1.23) oscillates very rapidly in the semiclassical limit, so that the dominant asymptotic contributions to the integral tend to arise from *stationary-phase points* contained within the domain of integration (Olver 1974), i.e., from points such that $d\phi_\pm/d\lambda = 0$. From (1.18)–(1.20), we find that

$$2\,d\eta_\lambda^{\text{WKB}}/d\lambda = \Theta(\lambda) \qquad (1.26)$$

where $\Theta(\lambda)$ is the classical deflection angle (1.3) associated with the angular momentum $\lambda$. Thus, by (1.25), a stationary-phase point $\bar{\lambda}$ corresponds to

$$\Theta(\bar{\lambda}) = \pm\theta \qquad (1.27)$$

i.e., to the relation (1.4) for $n = 0$. The sign is positive or negative depending on whether the path is repulsive or attractive.

By applying the method of stationary phase to the asymptotic evaluation of the integral (1.23) (Berry 1969), we find that, if only one stationary-phase point contributes,

$$f(k,\theta) \approx (ik)^{-1} \zeta_1 \zeta_2 \left[ \frac{\overline{\lambda}}{\sin\theta \, |\theta'(\overline{\lambda})|} \right]^{1/2} \exp\left\{ i\left[ 2\eta_\lambda^{\text{WKB}}(\overline{\lambda}) \pm \overline{\lambda}\theta \right] \right\} \quad (1.28)$$

where the phase factors $\zeta_1$ and $\zeta_2$ are $\exp(\pm i\pi/4)$ depending on whether the path is repulsive or attractive and on the sign of $\theta'(\overline{\lambda})$, respectively. The expression within the square brackets in the exponential is just $S(\theta)/\hbar$, where $S(\theta)$ is the *classical action* along the path (Berry & Mount 1972). The stationary-phase path thus verifies the action principle, as a classical path should.

The differential cross section $|f(k,\theta)|^2$ derived from (1.28) is readily seen to coincide with the classical differential cross section (1.5), completing the transition to classical mechanics in this case.

When more than one stationary-phase point (classical path) contributes, the amplitude becomes a sum of contributions of the form (1.28), provided that the stationary-phase points are of the first order and well-separated from one another. However, the resulting expression for $|f(k,\theta)|^2$ will also contain *interference terms*. This is a notable difference between classical and semiclassical results (Ford & Wheeler 1959).

The interference terms tend to be rapidly oscillatory, falling below the limit of resolution of ordinary macroscopic detectors, so that the last stage in the transition to the classical result (1.5) arises from the averaging over angles associated with the macroscopic detection process that washes out the interference oscillations.

The expression for the scattering amplitude as a sum of contributions of the form (1.28), associated with all classical paths leading to the direction $\theta$, is sometimes called the 'primitive' (or 'crude') semiclassical approximation.

## 1.3  Critical effects

The primitive semiclassical approximation derived in Section 1.2 goes beyond classical mechanics by including the effects of interference among contributions from different classical paths. Besides interference, the transition from geometrical to wave optics brings in *diffraction effects*.

Diffraction is most prominent where the deviations from the basic assumptions of geometrical optics (smooth and slow variation within a wavelength) are largest. This happens, e.g., at focal points or lines, or, more generally, at *caustics*. A caustic (literally, 'burning curve') is an envelope of a family of rays; familiar examples are the bright lines one can see on the inside walls of a teacup illuminated by sunlight. Since there is an infinite concentration of rays at a caustic, it corresponds to an infinite intensity according to geometrical optics. Focal points and lines are particular cases.

Diffraction effects are also expected to be large where geometrical optics predicts a finite discontinuity in the intensity, e.g., at a light/shadow boundary. Characteristically, the magnitude of the effects increases with the wavelength.

The singularities in classical scattering discussed in Section 1.1 give rise to strong diffraction effects. In the context of semiclassical scattering, we will call them *critical effects*.

(I) *Rainbow scattering*. We see from (1.28) that the primitive semiclassical approximation to $f(k,\theta)$ diverges at a rainbow angle, where $\theta'(\bar{\lambda}) = 0$. From (1.25)–(1.26), it follows that the assumption of first-order stationary-phase points breaks down at a rainbow angle. As will be seen in Chapter 3, a rainbow corresponds to a *caustic direction*. On one side of a rainbow angle, one also finds two stationary-phase points close together, as is illustrated by paths 4 and 6 in fig. 1.2.

(II) *Glory scattering*. If (1.28) could be extrapolated to the forward or backward directions, $f(k,\theta)$ would diverge like $(\sin\theta)^{-1/2}$ for a path with $\theta(\bar{\lambda}) = 0$ or $\pi$, $\bar{\lambda} \neq 0$. As will be seen in Chapter 4, forward or backward glory scattering is associated with a *focal line* along the axis. As may be seen in fig. 1.2(*b*), near a glory path, two stationary-phase points also come close together (corresponding to the repulsive path 2 and the attractive path 4 in this figure).

(III) *Forward diffractive scattering*. For paths that get deflected to very small scattering angles [violating condition (1.12)], the primitive semiclassical approximation breaks down for several reasons: the term '1' within the square brackets in (1.14) can no longer be neglected; the approximation (1.21) is not valid; the phase shifts become small, so that one no longer has rapidly oscillating integrands and the method of stationary phase cannot be employed. Diffraction effects become dominant in this region.

(IV) *Orbiting and resonance scattering*. Close to a situation in which the classical orbiting conditions (1.8) are fulfilled, one has effective potentials with a pocket, as shown in fig. 1.3. Instead of the bounded oscillatory orbits within such a pocket found in classical mechanics, what this situation leads to in quantum mechanics is the possibility of sharp *resonances*. A particle that undergoes resonant scattering can get trapped for a long time within the potential, in analogy to a classical particle orbiting around the scattering center.

The primitive semiclassical approximation breaks down in these situations, and we need more powerful techniques to handle them. Diffraction effects must be taken into account. An accurate treatment of diffraction is a notoriously difficult problem in wave optics, and hard problems in quantum dynamics are also connected with diffraction.

The angular ranges where critical scattering effects occur tend to get narrower and narrower as the wavelength decreases, as is characteristic of diffraction, so that one might think that they do not play a very significant role. However, just the opposite is true: in these ranges, the differential cross section is rapidly varying and often attains very large values (corresponding to the divergence along the critical directions in the classical limit), so that they tend to be associated with very prominent features. Furthermore, as may be inferred from fig. 1.2, the corresponding paths tend to probe the potential at very short distances, revealing important features of the interaction.

In the pioneering paper by Ford & Wheeler (1959), the critical effects were treated at a level similar to that of classical diffraction theory, leading to *transitional asymptotic approximations* to the scattering amplitude in the critical angular domains. Such approximations are characterized by very narrow ranges of validity, and by not merging smoothly with the behavior of the amplitude in neighboring angular domains. One should look for *uniform asymptotic approximations* that are free from these defects. One also wants to have an estimate of the accuracy of the approximations, by comparing them with 'exact' results.

To do this, one needs some model that can be solved exactly, or at least with high accuracy. Nontrivial exactly soluble models in physics are very few. Fortunately, one of them provides just what is needed, exhibiting all of the critical scattering effects and allowing one to test the accuracy of various approximations. This is the problem of scattering by a *homogeneous sphere.* In quantum mechanics, this corresponds to a square potential well or barrier; in optics [cf. (1.9)], it corresponds to *Mie scattering* ( Born & Wolf 1959 ).

Some of the most beautiful natural optical phenomena arise from Mie scattering by water droplets in the atmosphere, providing striking visual manifestations of critical effects. Before turning to the theory of Mie scattering, we review these phenomena in the next few chapters, describing also the evolution of attempted theoretical explanations, which is closely linked with the development of theories about the nature of light. Intuitive pictures associated with these effects will serve as a useful background for the discussion of more refined approximation methods given in succeeding chapters.

# Diffraction and Coronae

*For in June 1692, I saw by reflexion in a Vessel of stagnating*
*Water three Halos, Crowns, or Rings of Colours about the*
*Sun, Like three little Rainbows, concentrick to his Body.*
(Newton 1704)

The first critical scattering effect we illustrate is forward diffractive scattering. After outlining the geometrical-optic description of the scattering of light by a homogeneous sphere, we give a brief account of the evolution of classical diffraction theory, and of its application to explain the coronae seen around the Sun and the Moon.

## 2.1  Geometrical optics

The simplest description of the propagation of light is given by the laws of geometrical optics. In a homogeneous medium, light propagates along straight lines (rays). Upon meeting a plane interface between two different homogeneous media, it gets partially reflected and partially transmitted: the paths of reflected and transmitted rays are determined by the laws of reflection and refraction, respectively. If the interface is curved, it is sufficient to replace it by its tangent plane at the point of incidence of the ray.

An illustration is provided by the path of a light ray meeting a homogeneous sphere, such as a water droplet in the atmosphere (fig. 2.1). The angles of incidence and refraction shown in the figure are related by Snell's law,

$$\sin \theta_1 = N \sin \theta_2$$

where $N$ denotes the refractive index of water relative to air in this example. An incident ray here gives rise to an infinite series of scattered rays: a directly reflected ray 0 (fig. 2.1), a directly transmitted ray 1, and a series (2, 3, ...) of rays transmitted after 1, 2, ... internal reflections.

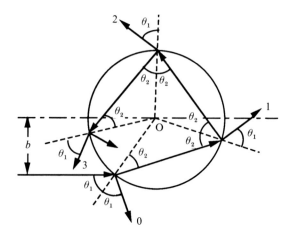

Fig. 2.1. Geometrical-optic ray tracing through a homogeneous sphere of refractive index
$N$. An incident ray with impact parameter $b$ produces an infinite series of scattered rays:
0 (directly reflected), 1 (directly transmitted), 2, 3, ... (transmitted after 1, 2, ... internal
reflections); $\theta_1$ = angle of incidence; $\theta_2$ = angle of refraction.

These laws suffice to determine the paths of light rays. How about
the propagation of the intensity? In geometrical optics, the energy flows
along the rays like an incompressible fluid, so that the intensity along a
thin conical pencil of light rays varies as the inverse of its cross-sectional
area (Born & Wolf 1959). Geometrical optics places no restrictions upon
the behavior of the intensity *transversely* to the light rays. For example,
when a parallel beam of rays meets an opaque sphere, a geometrical
*shadow* cylinder is formed behind the obstacle, and the intensity
undergoes a discontinuity at the light/shadow boundary.

At an interface between two different homogeneous media, the
intensity gets split between reflected and transmitted rays, but the
percentages of reflection and transmission are determined by laws that
are already outside the domain of geometrical optics. According to the
*Fresnel formulae* (Born & Wolf 1959), the percentages vary with the
angle of incidence and the polarization of the light.

For an air/water interface, most of the intensity gets transmitted,
except at very steep angles of incidence, so that high-order internal
reflections usually give a negligible contribution to the scattered intensity
in the example of fig. 2.1.

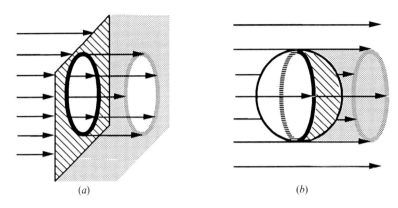

Fig. 2.2. *(a)* The edge of an aperture is its rim; *(b)* The edge of an obstacle is its shadow contour.

## 2.2 Classical diffraction theory

Diffraction in the broad sense is any deviation from the laws of geometrical optics. The diffraction of light from a point source by a small object was described in a book by Grimaldi published as early as 1665. Some of the light gets deflected into the geometrical shadow region, and the shadow is bordered by bright fringes.

While the laws of geometrical optics are consistent with a corpuscular theory of light, diffraction is a typical wave property. In 1804, Thomas Young proposed a theory in which diffraction is explained as a *local* effect, due to *edge rays*. An edge ray is one incident on the edge of an obstacle, defined as its intersection with the geometrical shadow boundary. As illustrated in fig. 2.2, the edge of an aperture in an opaque screen is the boundary of the aperture, while the edge of an obstacle is its shadow contour: edge rays are tangential to the obstacle.

According to Young's theory, an incident edge ray would undergo 'a kind of reflection', giving rise to what was later called a *boundary diffraction wave*. The propagation of this wave into the geometrical shadow would account for the penetration of light into this region; in the illuminated region, it would be superimposed onto the incident wave [fig. 2.3*(a)*], allowing one to explain the diffraction fringes bordering the shadow in terms of Young's principle of interference.

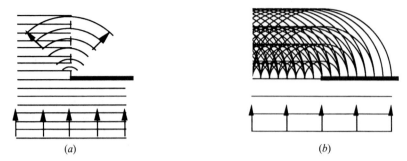

Fig. 2.3. *(a)* Young's boundary diffraction wave in diffraction by a half-plane.
*(b)* Diffraction by a half-plane according to the Huygens–Fresnel principle.

An alternative theory formulated by Fresnel in 1816 combined interference with *Huygens' principle*, according to which the propagation of a wave can be described in terms of secondary spherical waves emanating from every point of a wavefront. While Huygens employed this idea for a geometrical construction of succeeding wavefronts as the envelopes of his wavelets, Fresnel treated wave propagation as resulting from the interference of all secondary waves.

In terms of this *Huygens–Fresnel principle*, diffraction is explained as resulting from the perturbation of the interference pattern due to the blocking of a portion of an incident wavefront by the obstacle, i.e., as a *blocking effect*. This is illustrated in fig. 2.3*(b)* for the diffraction of a plane wave by an opaque half-plane: secondary waves arise only from the portion of an incident plane front that is not obstructed by the screen.

Fresnel submitted his memoir to compete for a prize offered by the French Academy of Sciences for an explanation of diffraction. Poisson, who was a member of the committee, derived from Fresnel's theory the prediction of a bright spot in the center of the shadow of a circular disc, and raised this as an objection, regarding it as an absurd result. Arago and Fresnel demonstrated by experiment that the spot really exists, so that Fresnel won the prize, but the effect became known as Poisson's bright spot! As will be seen later, it is related to the forward diffraction peak, as well as to forward glory scattering.

During the second half of the nineteenth century, an exact formulation of the Huygens–Fresnel principle was given by Helmholtz and Kirchhoff (Baker & Copson 1950, Born & Wolf 1959). Thus, for a scalar monochromatic plane wave perpendicularly incident upon an aperture in a plane screen, it can be shown that the scattering (diffraction) amplitude in a direction $\theta$ is *exactly* given by (Bouwkamp 1954)

$$f(k,\hat{s}) = -\frac{ik}{2\pi}\cos\theta \int u(x')\exp(-ik\hat{s}\cdot x')\mathrm{d}^2 x' \qquad (2.1)$$

where $\hat{s}$ is a unit vector in the direction of observation, the integration is performed over the plane of the screen and $u$ denotes the wave function on this plane.

Unfortunately, the exact boundary value of $u$ is unknown. Classical diffraction theory is based upon *Kirchhoff's approximation*, in which this unknown aperture distribution is replaced by a geometrical-optic one: $u$ is taken to coincide with the unperturbed incident wave over the aperture and it is taken equal to zero over the screen, on the shadow side. This leads to a well-defined formulation of the Huygens–Fresnel principle.

In spite of the crudeness of Kirchhoff's approximation, classical diffraction theory has led to reasonable results for the diffraction patterns of apertures with dimensions much larger than the wavelength, for small diffraction angles $\theta$, where most of the intensity is concentrated. For example, applying it to a circular aperture of radius $a$, one readily finds that (2.1), with Kirchhoff's approximation for $u$, yields

$$f(k,\theta)/a = -\mathrm{i}\cos\theta J_1(\beta\sin\theta)/\sin\theta \qquad (2.2)$$

where $J_1$ is the Bessel function of order one and $\beta = ka$ is the size parameter defined in (1.13). The argument of the Bessel function is just the path difference between rays in the direction $\theta$ coming from the center or the rim of the aperture. The intensity distribution associated with (2.2), known as the *Airy diffraction pattern*, represents a bright central circular spot surrounded by much weaker circular diffraction rings. The angular opening of the central spot, which gives the width of the *forward diffraction peak*, is given approximately by

$$\Delta\theta \approx 0.61\lambda/a \qquad (2.3)$$

an important result for the theory of the resolving power of optical instruments. This defines the domain of *small diffraction angles*: we see that they are of the order of $1/\beta$, the inverse of the size parameter.

Why classical diffraction theory works reasonably well in this domain can be understood from (2.1) (Nussenzveig 1959). It tells us that the amplitude for picking up the transverse wave number associated with the direction of scattering is proportional to the spatial Fourier component of the exact aperture plane distribution corresponding to this transverse wave number. For small diffraction angles, we are only probing very rough features of this distribution, down to a scale of the order of the aperture size. Finer details of the distribution, that reflect more difficult aspects of the dynamics of diffraction, affect mainly the domain of larger diffraction angles, where the intensity is weaker (besides giving rise to smaller corrections within the forward peak).

In spite of the apparent differences between Fresnel's and Young's explanations of diffraction, it turns out that classical diffraction theory has a fully equivalent formulation in terms of Young's interpretation. Indeed, it can be shown (Baker & Copson 1950) that the Fresnel–Kirchhoff representation of the diffracted wave as a surface integral over the unobstructed portion of a wave front actually depends only on its boundary, the edge line. By applying a transformation due to Maggi, the result can be transformed into a line integral over the edge and interpreted in terms of Young's boundary diffraction wave (Rubinowicz 1917, 1965). It should be stressed that this equivalence holds within the framework of classical diffraction theory, i.e., assuming the validity of Kirchhoff's approximation.

One consequence of this result is that, in classical diffraction theory, diffraction can also be regarded as a local effect, arising from the edge rays. Another consequence is *Babinet's principle* (Sommerfeld 1954): since the diffraction pattern depends only on the edge, it is the same for two 'complementary' arrangements, for which the aperture in one of them is replaced by the screen in the other. In particular, it is the same for a circular disc and for a circular aperture of the same radius; for that matter, the same Airy pattern, associated with (2.2), also holds for diffraction by a sphere of radius $a$, according to classical diffraction theory. This applies to the *diffracted* wave: the direction of incidence, along which the incident wave still propagates to infinity for a disc or sphere, but not for a circular aperture in an infinite opaque screen, must be excluded.

## 2.3   The corona

One beautiful natural manifestation of a forward diffraction peak is the corona. It is seen most often around the Moon, viewed through mist or thin clouds. It takes the form of a circular disc of light surrounding the Moon, bluish white near the Moon, with a reddish brown rim. This disc, called the *aureole*, is sometimes seen surrounded by one or more concentric colored rings.

The main reason why coronae are not seen just as often around the Sun is that one avoids looking directly at the Sun, even through a cloud. The brightness can be suitably attenuated by looking at the Sun's reflection in a water puddle (see the quotation by Newton at the beginning of this chapter). For color photographs of coronae around the Moon and the Sun, see Greenler 1980.

The angular radii of aureoles around the Moon can be estimated by direct comparison with the angular diameter of the Moon, which is about 0.5°. They are usually a few times this diameter, varying mostly between 2° and 5°.

Although coronae can also be produced by clouds of ice crystals, we consider only their formation in mist or clouds consisting of water droplets. Since the droplets are on the average far apart and randomly distributed, they act as independent scatterers (Sommerfeld 1954, Tricker 1970), so that their contributions are superimposed incoherently: it is sufficient to add the scattered intensities (forward scattering is excluded in this argument).

If the size distribution of droplets in the cloud is fairly uniform, the single-droplet scattering pattern just gets magnified by a factor of the order of the number of contributing droplets. The diffraction pattern produced by a single spherical droplet, according to classical diffraction theory, is the same as the Airy pattern due to a circular disc of the same radius, obtained from (2.2). The aureole is then nothing but the central diffraction peak, and its angular radius, according to (2.3), is directly proportional to the wavelength, thus explaining the reddish outer rim and the bluish tint near the Moon; the overlap of diffraction peaks of different wavelengths decreases the purity of the colors. From the quoted values of the usual range of variation of observed angular radii and from (2.3), we conclude that the size parameters of the droplets that are responsible for lunar coronae range roughly from about 50 to 100, corresponding to droplet radii of a few hundredths of a millimeter.

Actually, even for uniform droplets, this treatment is oversimplified, because one must also take into account, especially for the lower range of size parameters, the interference between the diffracted light and that transmitted through the droplets, which leads to an 'anomalous diffraction' pattern (van de Hulst 1957). A nonuniform size distribution of droplets in the clouds distorts the diffraction pattern: rings of different colors become irregular curves, associated with level lines of the size distribution, giving rise to the phenomenon of *cloud iridescence* (Greenler 1980, Minnaert 1954, Tricker 1970).

# 3

# The rainbow

*The rainbow is such a remarkable marvel of*
*nature ...that I could hardly choose a better suited*
*example for the application of my method.*
(Descartes 1637)

The main theories that have been proposed to explain the observed features of the rainbow, from Descartes to the present, are reviewed here. They evolved in close parallel with theories about the nature of light. Surprisingly, as late as 1957, it was pointed out that no reliable quantitative theory of the meteorological rainbow existed at that time (apart from complicated numerical computer calculations).

## 3.1   Geometrical-optic theory

In the rainbow (for color pictures, see Greenler 1980), the *primary bow* is surrounded by the much fainter *secondary bow*, in which the colors appear in reverse order. The bows are just *sets of directions* in the sky, centered around the *antisolar point*, the direction opposite to that of the Sun: the higher the Sun's elevation, the lower the antisolar point below the horizon, and the smaller the portion of the rainbow arcs seen.

Each of the bows has a *bright side* and a *shadow side*, and the shadow sides face each other, so that the sky between the bows looks darker, as was described by Alexander of Aphrodisias around 200 AD (*Alexander's dark band*).

By means of an experiment performed at the remarkably early date of 1304, it was demonstrated by Theodoric of Freiberg that the rainbow arises from the scattering of sunlight by individual water droplets in the atmosphere. Employing a spherical flask filled with water to simulate a giant water drop, he correctly identified the paths of light rays that are responsible for both bows: for the primary bow, those that undergo one internal reflection (rays of class 2 in fig. 2.1), and for the secondary bow those that undergo two internal reflections (rays of class 3 in fig. 2.1).

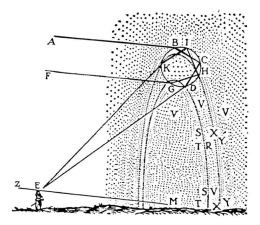

Fig. 3.1. Descartes' illustration of the formation of the rainbow: rays such as ABCDE form the primary bow, and FGHIKE the secondary one (from Boyer 1987, p. 209. Copyright © 1987 by Princeton University Press; reprinted by permission).

The mean opening angle of the primary rainbow cone (with respect to the antisolar direction) is about 42°; the supplement of this angle, about 138°, is the primary *rainbow scattering angle*. The mean scattering angle for the secondary bow is about 128°.

Theodoric's findings lay almost forgotten for three centuries, until they were independently rediscovered (by a similar method) by Descartes. In *Les Météores*, an appendix to his *Discours de la Méthode* (1637), he illustrated the Cartesian method by applying it to the rainbow (fig. 3.1). His main contribution was the explanation of a point that Theodoric had not discussed: scattered rays of classes 2 and 3 in fig. 2.1 emerge over broad ranges of scattering angles, corresponding to the continuum of impact parameters of the incident solar rays. Why then are the rainbow scattering angles singled out?

To answer this question, Descartes painstakingly traced the paths of a large number of incident rays through the droplet, applying the laws of reflection and refraction. He concluded that 'after one reflection and two refractions, there are many more rays that can be seen at an angle of from 41 to 42 degrees than at any smaller angle and there are none that can be seen at larger angles'. In modern language, he was plotting the deflection function for rays of class 2 and found the primary rainbow scattering angle to be a minimum.

Fig. 3.2 is a plot of the deflection function for rays of classes 2 and 3.

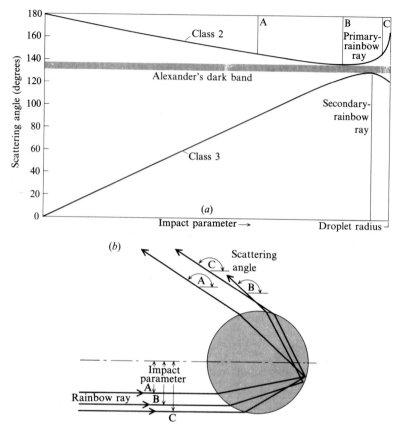

Fig. 3.2. *(a)* Scattering angle as a function of impact parameter for rays of classes 2 and 3. *(b)* On the bright side of the primary bow, two different class-2 rays A and C, with impact parameters on each side of the rainbow ray B, emerge at the same scattering angle (from Nussenzveig 1977. Copyright © by Scientific American, Inc. All rights reserved).

The secondary rainbow angle is a maximum for the deflection function of class-3 rays. The secondary bow is much fainter than the primary one because of the attenuation due to an extra internal reflection. Fig. 3.2 also shows that no rays of classes 2 and 3 emerge in the angular range between the two bows, thus explaining Alexander's dark band.

The association of rainbow scattering with an extremal of the deflection function agrees with the definition given in Chapter 1 [cf. (1.6)]. Rainbow angles are caustic directions at infinity, giving rise to an infinite concentration of the light intensity according to geometrical optics. The determination of rainbow scattering angles from the laws of geometrical optics is a simple exercise in maxima and minima; for the primary rainbow angle $\theta_R$, the result was given by Huygens in 1652:

$$\cos(\theta_R/2) = N^{-2}\left[(4-N^2)/3\right]^{3/2} \tag{3.1}$$

where $N$ is the refractive index. For the tertiary bow, the mean scattering angle is about 40°, so that it cannot be seen not just because it is fainter, but also because of the background glare around the Sun.

The explanation of the rainbow colors was given by Newton, following his prism experiments in 1666: the rainbow angle depends only on the refractive index [cf. (3.1)], and water is a dispersive medium. Newton also computed the width of the rainbow arcs by determining the rainbow angles for red and for violet light, taking into account the angular diameter of the Sun, and he checked the results against his own careful experimental observations. Additional information on the historical background may be found in Nussenzveig 1977 and Boyer 1987.

## 3.2   Wave-optic theory

Geometrical optics, which is consistent with a corpuscular theory of light, seems to account for the observed features of the rainbow. However, it does not explain one feature that is less often seen, the *supernumerary arcs*. One or more such arcs may appear just inside the primary bow (even more rarely, outside the secondary bow): their colors usually alternate between pale violet-pink and pale bluish-green (for a color picture, see Fraser 1983).

In order to explain the supernumerary arcs, it was necessary to employ a new theory about the nature of light, the wave theory. Young's explanation, given in 1804, is one of the finest natural illustrations of his principle of interference. On the bright side of the primary bow, where the supernumeraries appear, it may be seen in fig. 3.2 that there are two different rays of class 2, with impact parameters respectively below and above that of the Descartes rainbow ray, emerging in the same direction, with different optical paths. This is a general feature of rainbow scattering, as was noted in fig. 1.2.

According to geometrical optics, one should simply add the intensities associated with the two paths. However, light waves interfere: in the wave theory, the amplitudes are added, and the optical path difference between the two rays, which varies with the direction of observation, leads to constructive and destructive interference effects, giving rise to supernumerary peaks and valleys for each monochromatic component of the incident light.

In contrast with the position of the rainbow angle, the angular positions of the supernumeraries depend on the droplet size, which affects the optical path difference: the larger the droplet, the narrower is the angular separation of maxima and minima. Taking into account that the overlap of different colors tends to wash out the effect, one can understand that these arcs will only appear under favorable conditions of droplet size and uniformity.

Although Young's theory gave a qualitative explanation of the supernumerary arcs, it was like the primitive semiclassical approximation described in Sec. 1.2, in the sense that the interfering amplitudes were still constructed according to geometrical optics. Thus, the intensity would still be divergent at the rainbow angle, and it would be identically zero, for rays of a given class, on the shadow side of the rainbow.

These features are inconsistent with a true wave theory, in which abrupt shadow boundary discontinuities get smoothed over by diffraction. Diffraction is also required to explain the most difficult feature of the rainbow problem: the great variability in the appearance of the rainbow, depending on the size distribution of the droplets that produce it.

For large droplets, with radii of the order of a millimeter, one tends to see the whole range of bright prismatic colors; as the radius decreases, so does the brightness and purity of the colors, and some of the colors, e.g. the red, become weak or even disappear. Finally, for very small droplets found in clouds or fog, with radii of a few hundredths of a millimeter, one sees only a white band of light: this is the 'white rainbow' or 'fog bow' (for a picture, see Greenler 1980). This size dependence is a clear indication that diffraction must be taken into account.

The classical diffraction theory of the rainbow resulted from the work of the Cambridge school of mathematical physics, beginning with a detailed study by Potter in 1835 of families of rays and wavefronts associated with the rainbow problem. As illustrated in fig. 3.3, when one traces the paths of a family of rays through a droplet, several *caustics* are found, both inside and outside the droplet. In particular, scattered rays of class 2 give rise to a two-sheeted caustic outside, with one real sheet and one virtual (for a more careful discussion, see McDonald 1963).

The geometrical rainbow direction is asymptotic to the real sheet of this caustic, so that it represents a *caustic direction*. The deflection angle

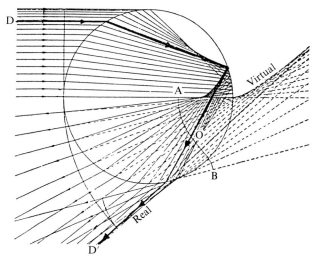

Fig. 3.3. Caustics formed by a family of incident rays; the class-2 ray caustic has a real and a virtual sheet. The Descartes rainbow ray DOD' emerges as the asymptote to this caustic. The S-shaped virtual wavefront AOB has an inflection point at O. A cusped wavefront outside with vertex on the caustic is shown (after Humphreys 1964).

of class-2 scattered rays increases with impact parameter up to the Descartes rainbow ray and then begins to decrease. A virtual wavefront inside the droplet (AOB, fig. 3.3) changes the sense of its concavity at the inflection point O. The double covering by rays on the bright side of the caustic, due to the propagation of this S-shaped wavefront, produces cusped wavefronts outside: the caustic is the locus of cusp vertices.

The wave-optical treatment of the rainbow therefore involves the diffraction of light in the neighborhood of a caustic. Potter's Cambridge colleague Airy applied classical diffraction theory to this problem in 1836. He chose Potter's S-shaped wavefront for the application of the Huygens–Fresnel principle; with origin at O (fig. 3.3), this wavefront is represented by a cubic equation near the origin. Approximating the unknown wave amplitude along this wavefront by a constant, Airy was led to introduce a new transcendental function to represent the distribution of light in the rainbow, his celebrated 'rainbow integral',

$$\mathrm{Ai}(z) = \left(3^{\frac{1}{3}}/\pi\right)\int_0^\infty \cos\!\left(t^3 + 3^{\frac{1}{3}}zt\right)\mathrm{d}t \tag{3.2}$$

known nowadays as the Airy function (Abramowitz & Stegun 1965).

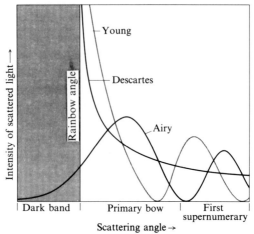

Fig. 3.4. Comparison of the rainbow theories of Descartes, Young and Airy (from Nussenzveig 1977. Copyright © by Scientific American, Inc. All rights reserved).

This may be regarded as a generalization of Fresnel's integral, with a qualitatively similar behavior: for negative $z$, corresponding to the bright side of the rainbow, it behaves like a slowly damped oscillation that has a decreasing period, representing the supernumerary peaks and valleys; on the dark side (positive $z$), it gets damped faster than exponentially, representing the penetration of light into the shadow by diffraction.

In fig. 3.4, Airy's prediction for the scattered intensity near the rainbow angle is compared with those of the Descartes and Young theories. Both of the earlier theories led to infinite intensity at the rainbow angle and zero on the shadow side, whereas Airy's theory yields a smooth transition between bright and dark sides, with the primary bow displaced toward the bright side of the rainbow angle.

Another significant difference between Young's and Airy's theories that is apparent in fig. 3.4 is the displacement in the predicted positions of supernumerary peaks: Young's theory predicts a valley where Airy's predicts a peak, and vice-versa. This arises from Young's neglect of an additional contribution to the phase difference between the two interfering rays: one of them goes through a focal line, and this gives rise to a phase difference of $\pi/2$ (van de Hulst 1957).

The angular scale of Airy's diffraction pattern increases as the droplet size decreases. The superposition of patterns arising from different monochromatic components of sunlight leads to quite different configurations, with varying degrees of overlap, thus accounting qualitatively for the changes in brightness, spectral distribution and purity of the rainbow colors with average droplet size. For very small droplets, the patterns become so broad that all colors overlap, giving rise to the 'white rainbow'. An attempt to correlate droplet size and rainbow appearance was made by Pernter (Pernter & Exner 1910), based upon Maxwell's theory of color vision.

## 3.3 Electromagnetic theory

It was verified by Brewster in 1812, following an earlier observation by Biot, that light from both the primary and the secondary bows is almost completely polarized. This can be readily verified by looking at a rainbow through Polaroid sunglasses and rotating them around the line of sight (for photographic evidence, see Können 1985).

The transverse character of light waves, responsible for their polarization properties, was one of the results explained by Maxwell's electromagnetic theory of light. After its formulation, it became possible to set up the problem of light scattering by a homogeneous sphere, such as a water droplet, as a well-defined boundary value problem in mathematical physics.

The exact solution of this boundary value problem in the form of a partial-wave series was given by Mie (1908). However, as is typical for such expansions [see the remarks following (1.17)], the numerical convergence of this series becomes very poor for large size parameters, requiring elaborate computer programming for its summation. This will be further discussed in Chapter 5.

A critical examination of the basic assumption underlying Airy's approximation was made by van de Hulst (1957). He concluded that it may be applied only if no rays with a deflection exceeding about 0.5° from the geometrical rainbow are involved, which limits the validity of Airy's theory to size parameters $\beta > 5000$, corresponding to droplet diameters larger than 1 millimeter. For such large values, droplet deformation to oblate spheroidal shape during free fall begins to be important, so that the model itself must be modified.

It was van de Hulst's 'somewhat disappointing conclusion' that a quantitative theory of the rainbow for the broad range of size parameters occurring in nature was still lacking, apart from numerical summations of the Mie series. In other words, only computers understood the rainbow!

# 4

# The glory

*And art thou nothing? Such thou art, as when*
*The woodman winding westward up the glen*
*At wintry dawn, where o' er the sheep track's maze*
*The viewless snow-mist weaves a glist'ning haze,*
*Sees full before him, gliding without tread,*
*An image with a glory round its head;*
*The enamoured rustic worships its fair hues,*
*Nor knows he makes the shadow, he pursues!*
(Coleridge, *Constancy to an Ideal Object* )

We start by reviewing the history of glory observations and their consequences, which is almost as remarkable as the phenomenon itself. We describe various proposed explanations of the effect: until recently, the best one available was based on a conjecture by van de Hulst, related with the geometrical theory of diffraction, which is also briefly discussed.

## 4.1    Observations

The earliest recorded observations of the glory were made between 1737 and 1739 (cf. Lynch & Futterman 1991), during a French geodetic expedition to Peru, undertaken to settle the dispute between Newton and Cassini on the figure of the Earth (occasioning Voltaire's famous quip: *'Vous avez confirmé dans des lieux pleins d'ennui, Ce que Newton connut sans sortir de chez lui'*). Early one morning, a few members of the expedition, including Bouguer and a Spanish captain, Antonio de Ulloa (who introduced platinum into Europe following this expedition), stood on top of Mount Pambamarca, in the Peruvian Andes.

What they saw was described by Bouguer as (Pernter & Exner 1910)

a phenomenon that must be as old as the world, but which no one seems to have
observed so far... A cloud that covered us dissolved, letting through the rays of the
rising sun... Then each of us saw his shadow projected upon the cloud... The
closeness of the shadow allowed all its parts to be distinguished: arms, legs, the head.
What seemed most remarkable to us was the appearance of a halo or glory around the

head, consisting of three or four small concentric circles, very brightly colored, each with the same colors as the primary rainbow, with red outermost.

Ulloa gave a similar description (Juan & Ulloa 1748), adding:

The most surprising thing was that, of the six or seven people that were present, each one saw the phenomenon only around the shadow of his own head, seeing nothing around other people's heads.

How remarkable the effect must have looked can be verified in recent color pictures of the glory around the shadow of the photographer's camera projected on mist (Brandt 1968, Greenler 1980).

Many observations of the glory were made during the nineteenth century from the top of the Brocken mountain in Germany, so that it also became known as 'the specter of the Brocken'. In 1933, during a mountaineering trip in which Heisenberg and Bohr took part, they saw it. As related by Heisenberg (1971),

Niels, who seemed particularly delighted by this unusual spectacle, said that he had heard about it before. People had told him that it was possibly the origin of the halo in which the early masters had wreathed the heads of the saints. 'And perhaps it is characteristic', he added with a wink, 'that it is only around our own heads that we can see the halo'.

Already in the sixteenth century, Cellini had been deeply impressed by observing a luminous white aureole around the shadow of his head on dewy grass, believing it (as he relates in his *Autobiography*) to be a mark of divine favour. What he saw was not the glory, but the *Heiligenschein*, that arises from the focusing properties of dewdrops: they project images of the sun on the grass blades behind them, which act like projection screens (Tricker 1970, Greenler 1980). On the other hand, the glory was a favorite image among nineteenth-century Romantic writers (Hayter 1973), as exemplified by Coleridge's beautiful poem quoted at the beginning of this chapter.

Other sightings of the glory during the nineteenth century were made from balloons, around the shadow of the balloon on the clouds. Today, it is commonly seen from airplanes, and is sometimes referred to as the pilots' bow. For color pictures of the glory around the shadow of airplanes, see Bryant & Jarmie 1974, Greenler 1980 and Können 1985.

In September 1894 I spent a few weeks in the Observatory which then existed on the summit of Ben Nevis, the highest of the Scottish hills. The wonderful optical phenomena shown when the sun shone on the clouds surrounding the hill-top, and especially the coloured rings surrounding the sun (coronas) or surrounding the shadow cast by the hill-top or observer on mist and cloud (glories), greatly excited my interest and made me wish to imitate them in the laboratory.

The writer of these lines did not succeed in his original intention, but the apparatus he designed for this purpose, the cloud chamber, revealed other interesting phenomena! The words are taken from C. T. R. Wilson's Nobel lecture of 1927 (Nobel Foundation 1965).

## 4.2   Proposed theories

The role of the shadow (of the observer's head, camera, plane, ...) in observing the glory is simply to define the antisolar direction: the glory is a *backscattering* effect. The whole effect is contained within a very narrow solid angle, with a typical opening of a few degrees, around the backscattering direction, thus 'each one is his/her own saint'.

The earliest attempted explanations of the glory were based upon a mistaken analogy with diffraction coronae. Since coronae are associated with forward diffraction, one needed a mechanism to reverse the direction of propagation of the scattered light. It was proposed by Fraunhofer and later by Pernter (1910) that light diffracted by the foremost droplets in a cloud (according to Fraunhofer, by droplets around the observer's head!) somehow got reflected by deeper portions of the cloud, so that the glory would be due to secondary scattering of the forward peak.

However, such proposals are untenable, for several reasons: 'regular reflection' from within the clouds does not arise; the observed angular distribution of the glory intensity is quite different from the Airy pattern (2.2) associated with coronae: it decays more slowly, and the central spot can be either brighter or darker than its surroundings, depending on droplet sizes (the pattern may vary during the course of an observation); the glory diffraction rings, in contrast with those of coronae, are strongly polarized.

It was shown by Ray (1923), by experimenting with artificial clouds, that the glory is a *primary* scattering effect, arising from backscattering by individual water droplets, i.e., already contained in the single scattering pattern. Ray tried to explain the glory in terms of interference between axial rays directly reflected from the droplet outer surface and those that undergo one internal reflection after transmission into the droplet. However, the resulting intensity would be far too small (cf. Chapter 11) and it would not yield the observed features. Similar remarks apply to a treatment by Bricard (1940).

Bucerius (1946) proposed an explanation in terms of a 'backwards

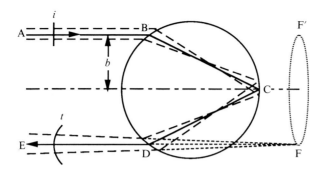

Fig. 4.1. An incident linear wavefront portion $i$ around a glory ray ABCDE emerges as a circular wavefront $t$ with a virtual focus at F. By axial symmetry, the locus of virtual foci is the ring source FF' that generates toroidal wavefronts, producing the axial focusing effect.

diffraction' effect, that would always lead to a backward dip (dark central spot). The results again disagree with the observed patterns of intensity and polarization.

A new mechanism that might account for the glory was suggested by van de Hulst (1947, 1957). To explain it, let us assume at first the existence of a *backward glory ray* such as ABCDE in fig. 4.1, i.e., a ray with nonzero impact parameter $b$ that emerges in the exactly backward direction [cf. (1.7)]. As is indicated in the figure by tracing the paths of neighboring rays, a portion of a plane incident wavefront emerges as a portion of a curved wavefront that appears to emanate from a virtual focus F.

For a generic scattering angle $\theta$, F would be a virtual point source, giving rise to spherical wavefronts. However, for $\theta = \pi$, in view of the axial symmetry, one sees, by rotating the whole figure around the axis, that one gets a whole focal circle, corresponding to a *virtual ring source* of radius $b$, that gives rise to *toroidal wavefronts*.

For scalar waves, the effect produced by this ring source along a direction $\theta$ very close to the backward one may readily be computed, e.g., by applying (2.1). The result is

$$f(k,\theta) \propto J_0(\lambda_G \sin\theta) \tag{4.1}$$

where $\lambda_G = kb$ is the glory angular momentum. For $\lambda_G(\pi - \theta) \gg 1$, we may apply the asymptotic expansion of the Bessel function, leading to the characteristic $(\sin\theta)^{-1/2}$ behavior found in (1.28). However, the axial caustic is suppressed by diffraction: (4.1) yields instead an amplitude enhancement by a factor of order $\lambda_G^{1/2}$ as compared with a generic (non-axial) direction. This arises from the constructive interference along the axis of contributions from all points of the virtual ring focus, so that we will call it the *axial focusing enhancement.*

This axial focusing effect might account for the backwards intensity enhancement seen in the glory. Unfortunately, however, a glory ray of the type illustrated in fig. 4.1 does not exist for the refractive index of about 1.33 which is characteristic of water. The largest deviation for rays of this class, attained at tangential incidence, leads to $\theta \approx 165°$, leaving a gap of about 15° to be bridged for emergence in the backward direction!

It was suggested by van de Hulst that this gap might be bridged by *surface waves* traveling along the surface of the sphere. Such waves are found in the *geometrical theory of diffraction*, which is briefly reviewed in the following section.

## 4.3   The geometrical theory of diffraction

This theory was proposed by Keller (1958, 1962) as a heuristic approach to diffraction, representing an extension of geometrical optics. It is based on the following postulates:

(i) The diffracted field propagates along *diffracted rays*. These rays are generated at the boundaries of diffracting objects in accordance with specific 'laws of diffraction', just as reflected or refracted rays are determined by the laws of reflection and refraction.

The paths of reflected or refracted rays obey Fermat's principle (Born & Wolf 1959): they are stationary in the class of all paths that touch the boundary between two media at one point, assumed to be an *interior* boundary point, i.e., not to lie on an edge or vertex. The paths of diffracted rays are determined by a generalization of Fermat's principle, extended to include points lying on edges or vertices, as well as arcs lying on boundaries.

For example, a ray incident on the edge of an aperture in a screen generates a cone of diffracted rays, making the same angle with the edge as the incident ray [fig. 4.2(a)]. In diffraction by a smooth obstacle [fig. 4.2 (b)], a point Q on an incident ray may be joined to a point P in the geometrical shadow by a diffracted ray having an arc on the boundary. By the extended Fermat principle, the diffracted ray follows the path that would be obtained by stretching a taut string over the surface, connecting

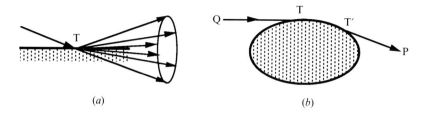

(a) (b)

Fig. 4.2. *(a)* Cone of diffracted rays generated by a ray incident on an edge at T. *(b)* Diffracted ray TT'P generated by the incident edge ray QT.

the points Q and P: straight line segments QT and T'P tangential to the obstacle at T and T', respectively, connected by a geodesic arc TT' along the surface.

(ii) Away from boundaries, amplitude and phase are transported along the diffracted rays according to the laws of geometrical optics.

(iii) The generation of diffracted rays is a *local effect*, depending only on the nature of the boundary and on the incident field in the immediate neighborhood of the point T where the diffracted ray is generated (fig. 4.2).

Thus, the initial value of the field on the diffracted ray is taken to be the incident field multiplied by a *diffraction coefficient*, which is determined by the local configuration around the point T. In diffraction by a smooth body, the surface diffracted ray keeps shedding energy tangentially away from the surface, along new diffracted rays generated at each point of its path [fig. 4.3*(a)*], so that its amplitude gets exponentially damped. The *decay exponents* associated with this damping are also assumed to be determined by local properties of the surface.

The surface diffracted rays are associated with *surface waves*, also referred to as *creeping modes* (Franz 1957). Their reality is strikingly demonstrated by schlieren photographs of the diffraction of sound pulses by solid bodies (Neubauer 1973).

For a transparent body, surface diffracted rays also undergo critical refraction, penetrating within it at the critical angle at each point of the surface. This gives rise to shortcuts through the body and to a variety of possible paths, as illustrated in fig. 4.3 for a sphere. Each time an internal ray meets the surface, it can either undergo reflection or reemerge tangentially, travelling an additional arc along the surface before the next critical refraction. All paths of the type shown in fig. 4.3 such that the total

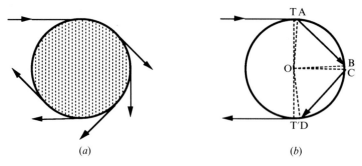

Fig. 4.3.*(a)* New diffracted rays are shed at each point along the path of a surface diffracted ray. The body surface is a caustic of diffracted rays. *(b)* Surface and critically refracted diffracted rays for a water droplet. The arc TA+BC+DT' spans an angle of ~15°.

angle described along the surface bridges the gap mentioned at the end of the previous section (about 15° for water) would contribute to the glory according to van de Hulst's conjecture.

Would the resulting amplitude, in spite of the surface wave damping, be large enough to account for the glory enhancement? To obtain quantitative results, one must know the values of the diffraction coefficients and decay exponents associated with the diffracted rays and their various interactions with the surface.

Unfortunately, these values cannot be determined within the strict context of the geometrical theory of diffraction. Since the amplitude propagates along the diffracted rays according to the laws of geometrical optics, it diverges at a caustic, and the body surface is a caustic of diffracted rays [fig. 4.3*(a)*] . Thus, the results of the geometrical theory cannot be applied at the surface.

In order to determine the values of the various coefficients, it was proposed by Keller (1958, 1962) that known solutions of certain 'canonical problems' be employed. By comparing asymptotic high-frequency expansions of these solutions with the 'Ansatz' supplied by the geometrical theory, expressions for the coefficients would be obtained.

For scattering by a transparent sphere, however, the canonical problem essentially coincides with the original one. Moreover, size parameters in the typical range for which the glory is observed are not sufficiently large for one to expect that reliable asymptotic estimates may be derived from the geometrical theory of diffraction.

Thus, at the time that van de Hulst formulated his conjecture, it could not be quantitatively tested, and the explanation of the glory remained an open problem.

# 5

# Mie solution and resonances

*It is nice to know that the computer understands
the problem, but I would like to understand it too.*
(Attributed to E. P. Wigner)

The Mie series solution for the scattering of an electromagnetic plane wave by a homogeneous sphere is introduced. Though 'exact', 'a mathematical difficulty develops which quite generally is a drawback of this "method of series development": for fairly large particles ... the series converge so slowly that they become practically useless' (Sommerfeld 1954). What is still worse, numerical studies reveal the occurrence in Mie cross sections of very rapid and complicated fluctuations, known as the 'ripple', that are extremely sensitive to small changes in the input parameters. They are related to orbiting and resonances, which may also be detected through their contribution to nonlinear optical effects, such as lasing, that are observed in liquid droplets.

## 5.1   The Mie solution

We consider a monochromatic linearly polarized plane electromagnetic wave with wave number $k$ incident on a homogeneous sphere of radius $a$ and a complex refractive index (relative to the surrounding medium) $N = n + i\kappa$, where $n$ is the real refractive index and $\kappa$ is the extinction coefficient (Jackson 1975). The time factor $\exp(-i\omega t)$, where $\omega = ck$ is the circular frequency, is omitted.

Taking the origin at the center of the sphere, the $z$ axis along the direction of propagation of the incident wave and the $x$ axis along its direction of polarization, the incident electric field is $\mathbf{E}_0 = \exp(ikz)\,\hat{\mathbf{x}}$. In spherical coordinates, the components of the scattered electric field at large distances are of the form (Bohren & Huffman 1983)

$$E_\phi \approx \frac{e^{ikr}}{ikr}\sin\phi\, S_1(\beta,\theta), \qquad E_\theta \approx \frac{e^{ikr}}{-ikr}\cos\phi\, S_2(\beta,\theta), \qquad (r \to \infty) \quad (5.1)$$

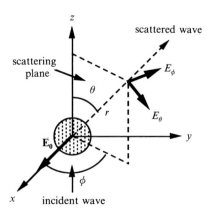

Fig. 5.1. Coordinate system and scattered electric field components.

where $\beta = ka$ is the size parameter [cf. (1.13)] and the complex quantities

$$S_j(\beta,\theta) \qquad\qquad (j = 1, 2)$$

are the *scattering amplitudes* associated with the two independent polarizations. In terms of the *scattering plane*, defined by the directions of incidence and of scattering (fig. 5.1), $S_1$ is associated with *perpendicular polarization* and $S_2$ with *parallel polarization*. The two amplitudes are elements of the *scattering matrix*, which is diagonal in the present case (van de Hulst 1957).

A complete set of scattering data consists of the *polarized intensities*,

$$i_j(\beta,\theta) \equiv \left| S_j(\beta,\theta) \right|^2 \qquad\qquad (j=1,2) \qquad\qquad (5.2)$$

and the *phase difference*

$$\delta \equiv \arg S_1 - \arg S_2 \qquad\qquad (5.3)$$

The scattered intensity in an arbitrary direction $(\theta,\phi)$ is given by (van de Hulst 1957)

$$i(\theta,\phi) = i_1 \sin^2 \phi + i_2 \cos^2 \phi$$

and the *scattering cross section* is

$$\sigma_{\text{sca}} = \pi k^{-2} \int_0^\pi [i_1(\beta,\theta) + i_2(\beta,\theta)] \sin\theta \, d\theta \tag{5.4}$$

The *degree of linear polarization* of the scattered radiation is

$$P \equiv (i_1 - i_2)/(i_1 + i_2) \tag{5.5}$$

The phase difference $\delta$ is required to treat the scattering of incident light with an arbitrary state of polarization (van de Hulst 1957).

The exact solution for $S_j$ was given by Mie (1908) [an earlier version is due to Lorenz (1890); for a historical account, see Logan 1965] in the form of a partial-wave expansion (Bohren & Huffman 1983):

$$S_j(\beta,\theta) = \tfrac{1}{2} \sum_{l=1}^{\infty} \left\{ \left[ 1 - S_l^{(j)}(\beta) \right] t_l(\cos\theta) + \left[ 1 - S_l^{(i)}(\beta) \right] p_l(\cos\theta) \right\}$$

$$(i, j = 1, 2; \ i \neq j) \tag{5.6}$$

The angular functions in (5.6) are defined by (Nussenzveig 1979)

$$p_\nu(x) \equiv \frac{P_{\nu-1}(x) - P_{\nu+1}(x)}{1 - x^2}, \qquad t_\nu(x) \equiv -x p_\nu(x) + (2\nu + 1) P_\nu(x) \tag{5.7}$$

where $P_\nu(x)$ is the Legendre function of the first kind, which becomes a Legendre polynomial when $\nu = l$ is an integer.

The functions $S_l^{(j)}(\beta)$ are $S$-matrix elements associated with magnetic ($j = 1$) and electric ($j = 2$) multipoles, respectively. They are given by

$$S_l^{(j)}(\beta) = -\frac{\zeta_l^{(2)}(\beta)}{\zeta_l^{(1)}(\beta)} \left[ \frac{\ln' \zeta_l^{(2)}(\beta) - Ne_j \ln' \psi_l(\alpha)}{\ln' \zeta_l^{(1)}(\beta) - Ne_j \ln' \psi_l(\alpha)} \right] \tag{5.8}$$

where

$$e_1 \equiv 1, \qquad e_2 \equiv N^{-2}, \qquad \alpha \equiv N\beta \tag{5.9}$$

and $\psi_l$ and $\zeta_l^{(1,2)}$ are the Ricatti-Bessel and Ricatti-Hankel functions, respectively defined in terms of spherical Bessel and Hankel functions by

$$\psi_l(z) = z j_l(z), \qquad \zeta_l^{(j)}(z) = z h_l^{(j)}(z)$$

## 5.2   Convergence difficulties

As has already been discussed following (1.17), the number of partial
waves that must be summed in an expansion such as (5.6) so as to get a
good approximation to the sum of the series, when $\beta \gg 1$, is roughly of
the order of the size parameter $\beta$. For scattering of visible light by a
sphere with 1 mm diameter, $\beta$ is of the order of 5000. Since each term of
the Mie series has a rather complicated structure, numerical studies of
Mie scattering require high-speed computers.

In view of the great practical importance of the problem,
considerable effort has been devoted to building efficient and reliable
programs. In a recent study by Wiscombe (1979, 1980), the series is cut
off after $l_{max}$ terms, where

$$l_{max} = \beta + 4\beta^{\frac{1}{3}} + 2 \qquad (5.10)$$

The reason why extra terms beyond $\beta$ are required will be discussed in
Chapter 8. In fact, one needs to go even beyond the cutoff (5.10) in order
to detect some extremely rapid fluctuations in Mie cross sections
(Chapter 14).

How smoothly does the Mie series converge? This can be
ascertained by plotting 'growth curves' of the intensity as a function of
the number of terms retained in the summation. The results (Bryant &
Cox 1966) show that the convergence is oscillatory and fairly irregular,
with a substantial contribution from the neighborhood of the upper limit
(5.10), particularly for large scattering angles.

However, the situation is much worse than might appear from the
above remarks. Fig. 5.2 shows a plot of the 'normalized backscatter
phase function', a suitably normalized version of the backscattered
intensity, for a water droplet, with absorption neglected ($N = 1.333 + 0i$)
and size parameters ranging from 4520 to 4524, a less than 0.1% variation
in radius. The results were computed from the Mie series, using double
precision arithmetic (11 significant digits) and a plotting interval of 0.01 in
$\beta$ (Shipley & Weinman 1978). According to (5.10), each plotted point
represents the sum of almost 5000 terms of the Mie series.

We see that, as the size parameter varies, the intensity undergoes
extremely rapid and complicated fluctuations . Note the logarithmic scale
in fig. 5.2: there are spikes showing an intensity change by over two

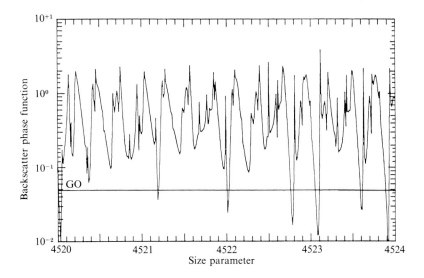

Fig. 5.2. Normalized backscatter phase function for $N = 1.333$ and size parameters near 4520. The straight line marked GO is the average geometrical-optic result (after Shipley & Weinman 1978).

orders of magnitude for a relative change in size parameter of order $10^{-5}$!

Similar fluctuations are seen in intensity plots at other angles, as well as in all kinds of Mie cross sections. This structure is known as the *ripple* (van de Hulst 1957, Kerker 1969, Bohren & Huffman 1983). The fluctuations are superimposed on a smoother background, with a relative amplitude that is maximal for backscattering and tends to decrease with the scattering angle; in the total cross section, they show up as a minor modulation.

The ripple is a highly sensitive function not only of the size parameter, but also of the complex refractive index $N = n + i\kappa$: small changes in any one of these parameters have drastic effects, reminiscent of 'sensitive dependence on initial conditions'. For water droplets found in the atmosphere, $\beta$ ranges from values of order unity to several thousand, $n$ remains in the range from 1 to 2, and $\kappa$ ranges from values of order $10^{-9}$ to values of order unity. Thus, although absorption has a smoothing effect, tending to remove the sharpest peaks, a sizable portion of the ripple can still be seen in domains of greater transparency.

One might think that the ripple is just a theoretical artifact, and that real liquid droplets are not homogeneous enough, or that they do not have a sharp and regular enough surface, for the Mie model to apply. This is not so: the ripple has been observed in a variety of experiments.

It has been seen in the backscattering of laser light from fiber-supported water droplets, by detecting the fluctuations in intensity as the droplet evaporates (Fahlen & Bryant 1968). It has been observed in radiation pressure measurements as well as in fixed-angle scattering from optically levitated liquid droplets (Ashkin & Dziedzic 1977, 1981), and comparisons of the measured features with Mie calculations have led to the most accurate determinations to date of absolute size and refractive index by light scattering (Chylek *et al* 1983).

The optical levitation experiments are a modern version of Millikan's oil drop experiment, in which the droplets are supported by the radiation pressure from a laser beam, rather than by a static electric field. The sharpness of the observed resonances has allowed detection of a fractional change in droplet radius $\delta a/a$ of 1 part in $10^6$, which, for $a = 10$ $\mu$m, corresponds to $\delta a \approx 0.1$ Å. Thus, the *average* surface of a microdroplet is defined to an accuracy of considerably less than a monolayer (Ashkin & Dziedzic 1977, Ashkin 1980)! This smoothness of the liquid surface on an almost atomic scale, as well as the nearly ideal spherical shape, result from the dominant role of surface tension. These observations show that the Mie model is in excellent agreement with reality.

Thus, the ripple and its sensitivity to parameter changes are inescapable features of Mie scattering. In practical applications, e.g., to radiative transfer in the atmosphere, one wants to average the results over a size distribution of droplets. In view of this sensitivity and of the enormous range of variation of the parameters in the atmosphere, numerical Mie computations of such averages for large size parameters, over extended intervals, become prohibitive even for supercomputers, as may be inferred from fig. 5.2.

It might be expected, on the other hand, that geometrical-optic (WKB) approximations would be valid, at least in this average sense, for large size parameters. The horizontal straight line in fig. 5.2 is the geometrical-optic prediction for the average over the plotted interval (the geometrical-optic curve is a sinusoidal oscillation about this average, with amplitude much smaller than that of the ripple fluctuations). We see that it is about one order of magnitude too small to account for the average Mie result. Thus, neither computers nor geometrical optics are the solution!

## 5.3 Lasing droplets

The high resolution employed in plotting fig. 5.2 and the large number of partial waves retained are still insufficient to reveal the full ripple structure that would be present within the plotting interval for this ideal model of a perfectly transparent sphere. As will be discussed in Chapter 14, this would require extending the summation to $l_{max} \approx N\beta$ rather than (5.10). The total number of peaks within the interval would exceed one thousand, and the narrowest ones would have widths of order ten to the minus a few hundred! In order to detect such needle-like peaks, one would have to retain a few hundred decimal places in the computation.

It will be seen in the later discussion that the narrow ripple peaks are associated with *resonances*. Radiation remains trapped for a long time within the sphere by traveling near the surface beyond critical internal incidence, so that it undergoes multiple nearly total internal reflection. This is a form of *orbiting*, a phenomenon that is closely related with resonances, as was noted in Sec. 1.3.

The assumption of perfect transparency that yields the needle-like resonances mentioned above is obviously unrealistic. However, several materials, including water, have transparency windows in the visible spectrum where absorption can be very low. While it is difficult to achieve the resolution required to detect extremely sharp resonances in scattering, the internal field intensity becomes very high at resonance, giving rise to observable *nonlinear optical effects.*

One may regard a microdroplet as an *optical cavity*, with a nearly ideal spherical shape maintained by surface tension, capable of supporting very high-$Q$ modes, rather analogous to *whispering-gallery modes* in acoustics (Rayleigh 1945, Walker 1978), through the optical feedback provided by almost total internal reflection. Optical effects associated with microdroplets are of great current interest (Barber & Chang 1988).

Stimulated emission into the resonant modes of solid spheres of $CaF_2$ doped with $Sm^{2+}$ having radii of the order of 1 mm was observed soon after the invention of the laser (Garrett *et al* 1961). Laser emission from dye-doped microdroplets, with radii of the order of 30 µm, has recently been observed (Tzeng *et al* 1984, Qian *et al* 1986). The operation is similar to that of a conventional dye laser, except for the optical feedback, which is provided by near-total internal reflection, rather than by external mirrors.

Color photographs of a stream of freely-falling ethanol droplets doped with Rhodamine 6G and irradiated by a single laser pulse of green light are reproduced in Qian *et al* 1986. Below a threshold intensity, one sees only a couple of bright green spots on each droplet, representing elastically scattered light; the origin of such glare spots will be discussed in Chapter 15. Above a threshold value of the incident intensity, the whole droplet circumference is highlighted in red, signalling that laser light emission is taking place, localized where the resonant internal field intensity is high, i.e., near the droplet boundary.

Other nonlinear optical effects have been seen in microdroplets. In particular, stimulated Raman scattering (SRS) from water droplets and from ethanol droplets with radii in the 20–50 μm range at relatively low excitation intensities has been observed (Snow *et al* 1985). The resonant internal field enhancement is so strong that it has allowed detection of up to 14th-order sequentially pumped SRS from $CCl_4$ droplets (Qian & Chang 1986); for further discussion, see Sec. 15.5.

These beautiful nonlinear optical effects provide direct visual evidence for the last of the four critical scattering effects listed in Chapter 1: orbiting and resonances. Together with the phenomena discussed in Chapters 2–4, they show that all four of these effects occur in Mie scattering. Since the exact Mie solution is available for numerical evaluation, the Mie model is ideally suited for investigating improved approximations to these effects, as well as their physical background.

# 6

# Complex angular momentum

> *The aim of a theory is to give a picture reproducing all
> the qualitative and quantitative features of the
> phenomenon considered. This aim is not attained until
> the solution obtained is of a sufficiently simple form.*
> (Fock 1948)

A vast amount of information on the diffraction effects that we want to study lies buried within the Mie solution. In order to understand and to obtain physical insight into these effects, as remarked by Fock in the above quotation, it is necessary to extract this information in a 'sufficiently simple form'. We now introduce the basic idea for doing this, the method of complex angular momentum, which goes back to Poincaré and Watson.

## 6.1 The Poisson representation

Our starting point is the Poisson sum formula (Titchmarsh 1937, Bremmer 1949, Schwartz 1966)

$$\sum_{l=0}^{\infty} \phi\left(l + \tfrac{1}{2}, x\right) = \sum_{m=-\infty}^{\infty} (-)^m \int_0^{\infty} \phi(\lambda, x) \exp(2im\pi\lambda) \, d\lambda \qquad (6.1)$$

where the 'interpolating function' $\phi(\lambda, x)$ reduces to $\phi(l+\tfrac{1}{2}, x)$ at the 'physical points' $\lambda = l+\tfrac{1}{2}$ ($l = 0, 1, 2, ...$), and $x$ denotes a set of parameters.

Applying this to the partial-wave expansion of the quantum scattering amplitude (1.14), we get the corresponding *Poisson representation*

$$f(k, \theta) = (ik)^{-1} \sum_{m=-\infty}^{\infty} (-)^m \int_0^{\infty} \lambda [S(\lambda, k) - 1] P_{\lambda - \frac{1}{2}}(\cos\theta)$$

$$\times \exp(2im\pi\lambda) \, d\lambda \qquad (6.2)$$

where $P_\nu(\cos\theta)$ is the Legendre function of the first kind and $\lambda$ (which will soon be extended to complex values) is called the *complex angular momentum* (in units of $\hbar$). The interpolating function $S(\lambda,k)$, which reduces to $S_l(k)$ at the physical points, is the extension of the $S$-function to – at first – continuous real values of $\lambda$.

Note that the $m = 0$ term in (6.2) is equivalent to the result of replacing the sum by an integral in the partial-wave expansion. A beautiful physical interpretation of the Poisson representation in the semiclassical limit was proposed by Berry & Mount (1972). In this limit, one can start from the WKB approximation to the $S$-function [cf. (1.18)] and, with the exclusion of near-forward and near-backward scattering, one can apply the approximation (1.21) for Legendre functions and omit the term '1' within the square brackets in (6.2) [cf. (1.22)].

With these approximations, (6.2) becomes

$$f(k,\theta) \approx (ik)^{-1}(2\pi\sin\theta)^{-\frac{1}{2}}$$

$$\times \sum_{m=-\infty}^{\infty}(-)^m \int_0^\infty d\lambda\,\sqrt{\lambda}\left[\exp\left(i\phi_{+,m}-i\frac{\pi}{4}\right)+\exp\left(i\phi_{-,m}+i\frac{\pi}{4}\right)\right] \quad (6.3)$$

where

$$\phi_{\pm,m}(\lambda,\theta) = 2\,\eta_\lambda^{\,\text{WKB}}+\lambda(2m\pi\pm\theta) \quad (6.4)$$

For $m = 0$, $\phi_{\pm,m}$ reduces to $\phi_\pm$, defined by (1.25), and we recover (1.23).

As we saw in Section 1.2, classical paths should be associated with stationary-phase points. Indeed, differentiating (6.4) with respect to $\lambda$ and employing (1.26), we find that a stationary-phase point $\bar{\lambda}_j$ corresponds to

$$\Theta\left(\bar{\lambda}_j\right) = -2m\pi \mp \theta \quad (6.5)$$

so that [cf. (1.4)] we complete the result found in (1.23), by now including the contributions from all classical paths (not just $m = 0$ ones).

At the same time, we obtain the physical interpretation of the integer $m$ in the Poisson representation: it is a 'topological quantum number' representing the *number of turns taken by a path around the origin*.

If we apply the method of stationary phase, we find that the contribution from a stationary-phase point has a form similar to (1.28),

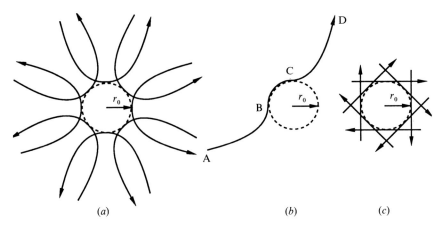

Fig. 6.1. (*a*) The sphere $r = r_0(\lambda)$ is the envelope of the family of classical paths with angular momentum $\lambda$; (*b*) A pseudoclassical path ABCD; (*c*) Same as (*a*) for free particles.

with the exponent representing the *classical action along the associated classical path* (Berry 1969). The net result is that, when interference among contributions from different paths can be neglected, one recovers the classical differential cross section (1.5), now with all classical paths taken into account.

In (6.3), however, contributions from *all* values of $\lambda$ and $m$ are included – not just those associated with stationary-phase points. To explain Berry and Mount's interpretation of such contributions, we consider the family of all classical paths with angular momentum $\lambda$. Each such path has the same classical distance of closest approach (outermost radial turning point) $r_0(\lambda)$. As illustrated in fig. 6.1(*a*), the sphere $r = r_0(\lambda)$ is the envelope of this family of classical paths.

It follows that an arc described on the surface of this sphere, though it is not a true classical path, is a *singular solution* of the equations of motion (Ince 1956). Fig. 6.1(*b*) illustrates what Berry and Mount call a *pseudoclassical path*: it is composed of an incoming classical path AB joined to an outgoing classical path CD (both with angular momentum $\lambda$) by an arc BC described on the caustic sphere $r = r_0(\lambda)$.

If one now computes the classical action along such a pseudoclassical path, where the arc BC corresponds to a total deflection angle $\Theta = -2m\pi \mp \theta$, the result coincides with (6.4): a particular case, for a true classical path, is the phase at a stationary-phase point $\bar{\lambda}_j$. We find, therefore, that a particle with *any* $\lambda$ can undergo *any* deflection $\Theta$ by following a pseudoclassical path, leading to the physical interpretation of the Poisson representation: it represents the scattering amplitude $f(k,\theta)$ as a *sum over all pseudoclassical paths that emerge in the direction* $\theta$. In this sense, we may regard (6.2) as a generalized version of Huygens' principle, or as a simplified form of Feynman's sum over paths (Feynman & Hibbs 1965).

For free particles, as indicated in fig. 6.1($c$), the classical paths are straight lines, and $r_0(\lambda) = b(\lambda) = \lambda/k$ is the impact parameter. The caustic sphere appears in the Debye asymptotic expansion for the cylindrical functions (equivalent to the WKB approximation) in the corresponding partial-wave component of an incident plane wave, providing a wave picture interpretation of the localization principle (Bremmer 1949). We also see, by comparison with fig. 4.3, that the diffracted rays of the geometrical theory of diffraction are a special type of pseudoclassical path: *pseudoclassical paths are a generalization of the concept of diffracted rays.*

As observed by Berry & Mount (1972), the Poisson representation (6.2) stands midway between classical and quantum mechanics: the 'topological sum' over $m$ leads back to the partial-wave expansion and quantized angular momentum, while summation over stationary action paths (averaging out interference effects) leads to classical mechanics.

We will adopt this vantage point to draw on both classical and quantum pictures in order to get insight into the critical effects. Thus, besides the semiclassical approach to quantum scattering, we will see that it may be advantageous to consider also a 'semiquantum approach' to classical scattering.

## 6.2  CAM approximations

The Poisson representation (6.2) is exact. In order to develop its potential for generating asymptotic approximations, the crucial step is to go over to the *complex $\lambda$ plane*. The basic ideas of the complex angular momentum method (from now on we will often use the abbreviation CAM to denote Complex Angular Momentum) go back to Poincaré (1910) and Watson (1918). A historical survey is given by Logan (1965).

For the models that will be dealt with here, the existence of an interpolating function $S(\lambda,k)$ that may be analytically extended to complex

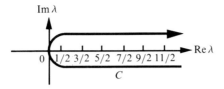

Fig. 6.2. Contour of integration $C$ for Watson's transformation

$\lambda$ is obvious by inspection [cf.(5.8)]. In a more general context, such as that of potential scattering, one may ask what physical properties of the interaction give rise to this analyticity. They may be shown to be related to the *finite range of the interaction* (Nussenzveig 1972a).

The original version of Watson's transformation was based on the formula

$$\sum_{l=0}^{\infty} \phi\left(l+\tfrac{1}{2},x\right) = \frac{1}{2} \int_C \phi(\lambda,x) \frac{\exp(-i\pi\lambda)}{\cos(\pi\lambda)} \, d\lambda \tag{6.6}$$

where $C$ denotes the contour shown in fig. 6.2. This can readily be checked by taking the residues of the integrand at the physical (half-integral) points. It is also easy to show the equivalence between this transformation and Poisson's sum formula (6.1) (Nussenzveig 1965), thus illustrating one of the possible ways in which the extension of the Poisson representation to the complex $\lambda$ plane may be made. However, the Poisson representation allows greater flexibility in the extension and its terms, as shown above, have a direct physical interpretation, so that it is a more convenient tool.

Once the extension to complex $\lambda$ has been made, one is free to deform the paths of integration in the Poisson representation around the $\lambda$ plane, taking due account of possible singularities. What are the advantages of this freedom?

In the partial-wave expansion, under semiclassical conditions, the contributions to the total scattering amplitude are distributed over a very large number of terms, with a complicated convergence pattern. The basic idea of CAM approximations is to concentrate the dominant asymptotic contributions to the amplitude (by taking advantage of the freedom of path deformation) within the neighborhood of a small number of points in the complex $\lambda$ plane, so as to obtain rapidly converging asymptotic expansions.

Watson's work addressed the problem of radio wave propagation around the Earth (for which the size parameter is several thousand), outside the line of sight of the transmitter, i.e., in the geometrical shadow region, treating the Earth as a perfectly conducting sphere. Applying (6.6) to the partial-wave expansion, he deformed the integral over the path $C$ onto an integral over the imaginary axis, which vanishes because the integrand is odd. In this process, the only singularities that are met are poles (the integrand is meromorphic), of which there is an infinite number, so that the result is transformed into a *residue series* over the poles.

Poles of the $S$-function in the $\lambda$ plane are now known as *Regge poles*, after the work of Regge (de Alfaro & Regge 1965), that has led to applications of complex angular momentum in high-energy physics. For a perfectly conducting sphere, the residues represent *surface wave* contributions, of the type illustrated in fig. 4.3. The imaginary parts of the poles are associated with angular damping (cf. Chapter 7), and they increase rapidly, so that the residue series converges very fast in the *deep shadow region*, thus achieving the aim of the transformation: it is sufficient to retain one or two terms.

Watson's conclusion was that the wave function gets exponentially damped as one penetrates into the deep shadow, and that this mechanism would not account for the observed effects of radio wave propagation around the Earth. The role of the ionosphere was still unknown at that time.

Within a geometrically illuminated region, the wave function can no longer be dominated by residue series contributions. One expects to find dominant asymptotic contributions from geometrical-optic paths, which we know to be associated with stationary-phase points on the real $\lambda$ axis [cf. (6.5)]. Thus, in the CAM method, *different transformations are required in different regions*.

Since a real stationary-phase point is also a saddle point, a good way to deform the path of integration in an illuminated region is to transform it into a *steepest-descent path* through the saddle point. This extension of the original Watson transformation was first proposed by White (1922). Generally, the contribution from a deformed path of integration in the $\lambda$ plane is known as a *background integral*.

These initial applications of the CAM method indicated that it should be a powerful tool for extracting asymptotic information from the partial-wave series. However, many problems remained open. Applying different transformations in different regions led to a matching problem across the boundaries between such regions. Since a typical case is a light/shadow boundary, it is precisely in connection with these transition domains that the most interesting diffraction phenomena, including the critical effects that we want to study, will occur.

Further extensions of the CAM method were required to deal with such problems. The results that have been achieved are described in the succeeding chapters.

# Scattering by an impenetrable sphere

*Thus, the way from the rigorous theoretical solution to the*
*approximate practical one took about 40 years of research.*
(Fock 1948)

We begin with a simple model, scalar (e.g., quantum) scattering by an impenetrable sphere. We review familiar high-frequency approximations to the solution: the WKB method and classical diffraction theory, as well as the geometrical theory of diffraction. These approaches fail within a neighborhood of the forward direction known as the *penumbra*. Fock's theory of what happens in this region, based on a picture of diffraction as transverse diffusion, yields the 'approximate practical solution' referred to in the above quotation. An early version of the CAM method confirmed and extended these results.

## 7.1 WKB and classical diffraction theory approximations

A quantum hard sphere, a highly idealized model for a strongly repulsive short-range interaction (e.g., in nuclei or in a 'hard-sphere gas'), is defined by the boundary condition that the Schrödinger wave function must vanish at its surface, $r = a$. Employing the partial-wave expansion of the solution outside the sphere,

$$\psi(r,\theta) = \sum_{l=0}^{\infty} \left(l + \tfrac{1}{2}\right) i^l \left[ h_l^{(2)}(kr) + S_l h_l^{(1)}(kr) \right] P_l(\cos\theta) \qquad (7.1)$$

(where $S_l \rightarrow 1$ for the incident plane wave), the boundary condition immediately yields the corresponding $S$-function:

$$S_l(\beta) = -h_l^{(2)}(\beta) / h_l^{(1)}(\beta) \qquad (7.2)$$

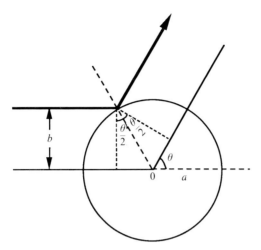

Fig. 7.1. The path difference between a ray geometrically reflected from the surface and one going through the center is $2a \sin(\theta/2)$.

In the WKB approximation (Messiah 1959, Schiff 1968), the phase of the wave function and the probability density behave, respectively, like the eikonal (optical path) and the intensity in geometrical optics. Thus, the sphere should scatter like a perfectly reflecting one would according to geometrical optics, i.e., isotropically.

As illustrated in fig. 7.1, geometrical-optic reflection from the surface leads to a phase factor $\exp[-2ika \sin(\theta/2)]$ in the scattering amplitude. Since the total cross section for geometrical reflection is $\pi a^2$, isotropy leads to $|f(k,\theta)|^2 = a^2/4$, so that the (zero-order) WKB approximation to $f(k,\theta)$ is determined up to a constant phase factor [cf. (1.28)].

Higher-order WKB corrections to the amplitude are obtained by assuming a power series expansion in the parameter $1/k$. In the present case, the result, including the first-order correction, is (Keller, Lewis & Seckler 1956)

$$f(k,\theta) = -\frac{a}{2}\exp\left[-2i\beta\sin(\theta/2)\right]\left[1 + \frac{i}{2\beta\sin^3(\theta/2)} + \cdots\right] \quad (7.3)$$

where the zero-order term agrees with the above considerations.

We would expect the WKB approximation to break down when the first-order correction becomes as large as the zero-order term. By (7.3), this happens for $\theta \lesssim \gamma$, where

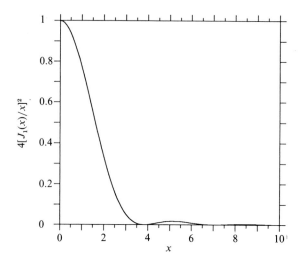

Fig. 7.2. The Airy diffraction pattern

$$\gamma \equiv \left(2/\beta\right)^{\frac{1}{3}} \tag{7.4}$$

is the *penumbra width* parameter, that sets the scale for all the effects connected with the penumbra. Thus, the domain of applicability of the WKB approximation is restricted to $\theta \gg \gamma$.

According to classical diffraction theory, as was mentioned at the end of Sec. 2.2, the diffraction pattern for the sphere is the same as that for an opaque circular disc with the same radius. While the diffracted *intensity* is also the same for a circular aperture in an opaque screen, the *amplitudes*, according to Babinet's principle (Sommerfeld 1954), are equal and opposite, so that, by (2.2), the diffraction contribution to the scattering amplitude for the sphere is

$$f(k,\theta) = \mathrm{i}a\, J_1\left(\beta \sin \theta\right)/\sin \theta \tag{7.5}$$

where we have set $\cos\theta = 1$, since (7.5) is to be applied only for $\theta \ll 1$.

The corresponding *Airy diffraction pattern* for the intensity, normalized to unity at $\theta = 0$, is plotted in fig. 7.2 as a function of $x \equiv \beta \sin\theta$. This is also the classical diffraction pattern associated with the corona (Sec. 2.3). About 84% of the diffracted intensity goes into the central peak, known as the *forward diffraction peak*. The angular width of this peak yields the expected domain of validity of the classical diffraction theory result, namely, $\theta \lesssim \beta^{-1}$ ($\ll \gamma$). Since the WKB approximation (7.3) is restricted to $\theta \gg \gamma$, this leaves an angular gap extending from $\theta \gg \beta^{-1}$ to the WKB domain, where neither one of these approximations should apply. This gap is the *penumbra region*.

According to these results, the angular distribution for $\beta \gg 1$ contains an isotropic component due to geometrical reflection and a highly anisotropic diffraction component, sharply forward-peaked and with much higher intensity, by a factor of order $\beta^2$, within this peak.

The total cross section follows from the *optical theorem* (Schiff 1968)

$$\sigma = (4\pi/k) \operatorname{Im} f(k,0) \tag{7.6}$$

Applying (7.5) in the forward direction, this leads to $\sigma = 2\pi a^2$, which is the well-known *extinction paradox* (van de Hulst 1957): in the short-wavelength limit, the total cross section becomes twice the geometrical cross section. The extra contribution is due to diffraction: the Airy diffraction pattern adds $\pi a^2$ to the cross section, as may be verified by integrating it over all solid angles (the solid angle associated with the forward peak is of order $\beta^{-2}$). More simply, by Babinet's principle, the diffraction cross section is the same as that of a circular aperture of radius $a$.

## 7.2  Geometrical theory of diffraction

According to Sec. 4.3 and fig. 4.3, diffracted rays in diffraction by an impenetrable sphere are excited by incident *edge rays*, i.e., those tangential to the sphere at points $T_1$ and $T_2$ in fig. 7.3. They generate 'creeping modes', that propagate along the surface of the sphere, shedding radiation away tangentially at each point along their path.

As was mentioned in Sec. 4.3, the amplitude for excitation of a given creeping mode $n$ by a tangentially incident ray is characterized by a *diffraction coefficient* $D_n$. By reciprocity, the same coefficient is associated with the shedding of tangentially outgoing rays by the surface waves. Thus, the surface wave contribution to the scattering amplitude is of the form (Levy & Keller 1959, Nussenzveig 1969a)

$$f_s(k,\theta) = -\frac{ia}{(\sin\theta)^{1/2}} \left\{ \sum_n D_n^2 \frac{\exp(i\lambda_n\theta)}{1+\exp(2i\pi\lambda_n)} \right.$$

$$\left. -i\sum_n D_n^2 \frac{\exp[i\lambda_n(2\pi-\theta)]}{1+\exp(2i\pi\lambda_n)} \right\} \tag{7.7}$$

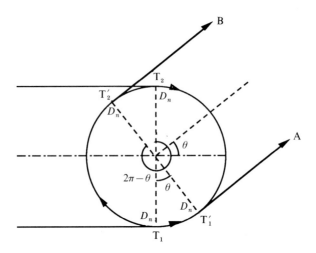

Fig. 7.3. Diffracted rays $T_1 T_1' A$ and $T_2 T_1 T_2' B$ in the direction $\theta$.

where $\lambda_n$ is the surface propagation constant of the $n$th mode; $\mathrm{Im}\lambda_n > 0$ represents the damping by radiation.

As illustrated in fig. 7.3, the two basic diffracted ray paths emerging in the direction $\theta$ are associated with surface angles $\theta$ and $2\pi - \theta$, for tangential incidence at $T_1$ and $T_2$, respectively. Once excited, the surface waves keep turning around the sphere, generating the factors

$$\left[1 + \exp(2i\pi\lambda_n)\right]^{-1} = \sum_{m=0}^{\infty} (-)^m \exp(2im\pi\lambda_n) \qquad (7.8)$$

The sign change at each turn, as well as the various powers of $(-i)$ in (7.7), arise from the phase change associated with the crossing of the axial focal lines; the factor $(\sin\theta)^{-1/2}$ is associated with the density of paths, as is the corresponding factor in (1.28).

As was explained in Sec. 4.3, in order to determine the values of the mode propagation exponents $\lambda_n$ and diffraction coefficients $D_n$, one has to look for known asymptotic solutions of 'canonical problems', outside the context of the geometrical theory of diffraction. In the present case, the canonical problem is precisely that with which we are dealing, scattering by an impenetrable sphere, and these parameters were determined by comparison with solutions obtained by the Watson transformation method (Franz 1957).

In particular, the exponents $\lambda_n$ are the Regge poles associated with this problem; as will be seen below (Sec. 7.3), we have

$$\left|\exp(i\lambda_n\theta)\right| \approx \exp\left(-\tfrac{1}{2}\sqrt{3}\,x_n\theta\;\gamma\right) \qquad (7.9)$$

where the numerical constants $x_n$ grow with $n$ and $\gamma$ is the parameter (7.4). It then follows from (7.7) and (7.9) that the surface wave contribution $f_s$ to the scattering amplitude is rapidly convergent (and becomes exponentially small) for $\theta \gg \gamma$. This is essentially equivalent to Watson's conclusion mentioned in Sec. 6.2.

Within the penumbra region, for not very large $\beta$, surface waves can give a non-negligible contribution, but there remains the problem of determining the dominant effects in this region. This was the problem addressed by Fock's theory of diffraction, which we now discuss.

### 7.3. Fock's theory of diffraction

Fock's impressive contributions to diffraction theory are gathered in his collected papers on this subject (Fock 1965). They are related to the parabolic equation method introduced by Leontovich (1944) and physically interpreted in terms of the concept of *transverse diffusion* of the wave amplitude along wavefronts (Malyuzhinets 1959). We give a brief review, based upon Malyuzhinets' presentation.

In geometrical optics, as was noted in Sec. 2.1, there are no constraints on the behavior of the amplitude transversely to the light rays: there is only a longitudinal transport equation for the amplitude (along the rays), and transverse discontinuities are allowed, giving rise to the singular behavior at shadow boundaries and caustics.

To see how the idea of transverse diffusion arises, let us begin with monochromatic wave propagation in free space, described by the Helmholtz wave equation

$$\left(\Delta + k^2\right)u(x,y,z) = 0 \qquad (7.10)$$

and let us substitute the 'Ansatz'

$$u(x,y,z) = A(x,y,z)\exp(ikz) \qquad (7.11)$$

to describe propagation in the $z$ direction with spatially variable amplitude. The result is

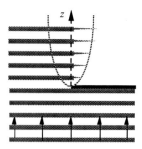

Fig. 7.4. Amplitude diffusion in diffraction by a half-plane (after Malyuzhinets 1959).

$$\left(1 - \frac{i}{2k}\frac{\partial}{\partial z}\right)\frac{\partial A}{\partial z} = \frac{i}{2k}\left(\frac{\partial^2 A}{\partial x^2} + \frac{\partial^2 A}{\partial y^2}\right)$$

In the semiclassical domain of high $k$, we can neglect the second term on the left side with respect to the first one. In a coordinate system moving along with a wavefront, $z = ct$, the above equation becomes

$$\frac{\partial A}{\partial t} = D\left(\frac{\partial^2 A}{\partial x^2} + \frac{\partial^2 A}{\partial y^2}\right) \qquad (7.12)$$

where

$$D = ic/(2k) \qquad (7.13)$$

The result (7.12) is then interpreted as a diffusion equation along the propagating wavefront, i.e., along directions transverse to the rays. To see how this process would account for diffraction, let us consider a plane wave of unit amplitude incident on an opaque half-plane at $z = 0$, with Kirchhoff boundary conditions on this plane: $A = 0$ on the screen (shadow side), $A = 1$ outside (illuminated region).

The boundary value problem for (7.12) then becomes analogous to the diffusion from an initial step-function profile, graphically represented in fig. 7.4 [to be compared with figs. 2.3(a) and (b)]. As the wavefront advances beyond the half-plane, the amplitude gradually diffuses along it, giving rise to Fresnel diffraction.

Fock applied this basic idea to diffraction by a smooth body, in particular, by a sphere. In this case, as shown in fig. 7.5, propagation into the shadow takes place through diffracted rays. The trace ABCD of a typical diffracted wavefront on the meridional plane is an involute of the circular section of the sphere (the curve traced by the tip of a taut piece of thread as it unrolls from the spool).

Fock's theory led to expressions for the wave function in some of the

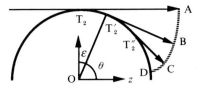

Fig. 7.5 Diffracted rays and corresponding wavefront ABCD in diffraction by a sphere.

penumbra regions in terms of new types of diffraction integrals known generally as Fock functions. One of the simplest examples is the result for the normal derivative of the wave function on the surface of a hard sphere (where the wave function vanishes). With the $z$ axis along the direction of incidence, $\varepsilon \equiv (\pi/2) - \theta$ measures the angular deviation from the shadow boundary on the surface (fig. 7.5), and Kirchhoff's approximation would yield

$$\chi(\beta,\varepsilon) \equiv \frac{1}{k}\left(\frac{\partial\psi}{\partial r}\right)_{r=a} = 0 \qquad (\varepsilon \geq 0)$$

$$= 2i\sin\varepsilon\exp(i\beta\sin\varepsilon) \qquad (\varepsilon \leq 0) \qquad (7.14)$$

where the first line represents the shadow region and the factor 2 in the second line arises from adding the contributions from the incident and geometrically reflected waves.

Fock's theory leads to the result (Fock 1965, Nussenzveig 1965)

$$\chi(\beta,\varepsilon) = \frac{e^{-i\pi/6}}{2\pi}\frac{\gamma\exp(i\beta\varepsilon)}{\sqrt{\cos\varepsilon}}\mathscr{F}(\varepsilon\,\gamma) \qquad (7.15)$$

where

$$\mathscr{F}(s) \equiv e^{i\pi/3}\int_{e^{2i\pi/3}\infty}^{e^{-i\pi/3}\infty}\frac{\exp\left(e^{-i\pi/6}sw\right)}{\mathrm{Ai}(w)}\,dw \qquad (7.16)$$

is one of Fock's functions, a sort of complex Fourier transform of the inverse of the Airy function (3.2). Note that the argument of this function in (7.15) is $\varepsilon/\gamma$, in agreement with the interpretation of $\gamma$ as the scale parameter for the angular width of the penumbra.

On the shadow side, when $\varepsilon \gg \gamma$, one can reduce the integral in (7.16) to a rapidly convergent series of residues at the poles, the zeros of the Airy function, which are located on the negative real axis. The result (Nussenzveig 1965) is a sum of surface wave contributions similar to (7.7), with

$$\lambda_n = \beta + e^{i\pi/3} x_n/\gamma, \qquad \mathrm{Ai}(-x_n) = 0 \tag{7.17}$$

i.e., $x_n$ is the $n$th zero of $\mathrm{Ai}(-x)$. For large $n$, $x_n$ grows like $n^{2/3}$, so that it suffices to retain the first few terms of the series, and (7.17) yields the dominant terms of the decay exponents in (7.7). Thus, deep within the shadow, the Fock function describes the exponential decay of the normal derivative on the surface (7.15).

On the illuminated side, for $|\varepsilon| \gg \gamma$, the integral (7.16) is dominated by a saddle point contribution, the evaluation of which (Nussenzveig 1965) leads to

$$\chi(\beta,\varepsilon) \approx 2i \frac{\varepsilon}{\sqrt{\cos\varepsilon}} \exp\left[i\beta\left(\varepsilon - \frac{\varepsilon^3}{3!}\right)\right], \quad |\varepsilon| \gg \gamma \tag{7.18}$$

This should match smoothly with the second line of (7.14), which can be shown to be the dominant term in this region. While the results agree to order $\beta\varepsilon^3$ in the exponent, the match is not quite smooth.

This illustrates a basic feature of the Fock approximation: it is a *transitional asymptotic approximation*, like the Airy theory of the rainbow (cf. also Sec. 1.3). It interpolates between the shadow and illuminated regions, describing diffraction effects within the penumbra, but it cannot be extended far enough into these regions to fully bridge the gap that separates them. Several examples of this behavior of the Fock theory will be given in Chapter 8.

The physical interpretation of the Fock theory in terms of transverse diffusion, though appealing, remains rather obscure, since the 'diffusion coefficient' (7.13) is purely imaginary, pointing towards an analogy with Schrödinger's equation, rather than the diffusion equation. The way that Airy functions appear in (7.16) and similar Fock functions also requires to be better understood. It will be seen in Chapter 8 that CAM theory provides the explanation of all these features and enables us to overcome the deficiencies of Fock's approximation.

## 7.4.  CAM theory

Versions of the CAM method available until 1965 could be applied only to disjoint and restricted spatial domains, with broad gaps among them and no systematic discussion of the transition regions where the critical diffraction effects take place. An improved version of the method (Nussenzveig 1965) led to the possibility of unrestricted extension to any domain in space; the first application was to discuss the behavior of the wave function in the hard sphere model both at finite distances and at infinity, though still employing transitional approximations.

The starting point is the Poisson sum formula (6.1) applied to the partial-wave expansion (7.1)–(7.2). There is, in principle, a considerable degree of arbitrariness in the choice of the interpolating function in the Poisson sum formula, since it has only to reproduce values at the discrete physical points. In practice, however, requirements of analyticity and good asymptotic behavior at infinity in the $\lambda$ plane lead to well-defined choices.

Nevertheless, different continuations are required depending on whether one wants to include the forward or the backward direction. This is because the Legendre function $P_\nu(x)$, while regular at $x = 1$, has a logarithmic singularity at $x = -1$. A simple example is provided by the CAM representation of the incident plane wave, given by

$$\exp(ikr\cos\theta) = \mp e^{\pm i\pi/4}\left(\pi/2kr\right)^{\frac{1}{2}}\int_0^{\sigma_\pm\infty} H_\lambda^{(1,2)}(kr)$$

$$\times \exp\left(\pm\tfrac{1}{2}i\pi\lambda\right)P_{\lambda-\frac{1}{2}}(\pm\cos\theta)\tan(\pi\lambda)d\lambda \quad (\theta \lessgtr \pi/2) \quad (7.19)$$

where upper and lower signs match each other and the limits and paths of integration are described in Nussenzveig 1965.

Thus, in the forward (backward) half-space, where the incident wave is outgoing (incoming), we have a representation in terms of outgoing (incoming) cylindrical waves, and of $P_{\lambda-\frac{1}{2}}(\cos\theta)$ $[P_{\lambda-\frac{1}{2}}(-\cos\theta)]$, which is regular at $\theta = 0$ ($\theta = \pi$), respectively. This illustrates the need for different representations in different angular domains.

After applying the Poisson sum formula with either a forward or a backward choice for the angular functions [using the relation $P_l(\cos\theta) = (-)^l P_l(-\cos\theta)$ in (7.1)], one can make use of reflection properties of the integrand functions to extend the integrals over the whole real axis of the $\lambda$ plane. This is followed by the crucial step of judiciously deforming the path of integration into the complex $\lambda$ plane, which requires knowledge of the analytic properties of the integrands and their asymptotic behavior at infinity along arbitrary complex directions.

The integrands turn out to be meromorphic functions of $\lambda$, and their singularities are the poles of the $S$-function, which, according to (7.2) and (6.1), is given by

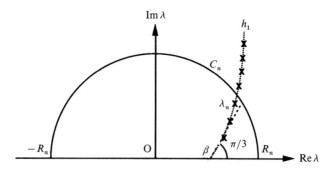

Fig. 7.6. Regge poles $\lambda_n$ (**x**) for a hard sphere, located on Stokes line $h_1$. The path $C_n$ passes halfway between consecutive poles.

$$S(\lambda,\beta) = -H_\lambda^{(2)}(\beta)\big/H_\lambda^{(1)}(\beta) \qquad (7.20)$$

These hard sphere Regge poles (cf. Sec. 6.2) are the zeros of the Hankel function in the denominator of (7.20). There is an infinite number of them, symmetrically located with respect to the origin.

The pole distribution in the upper half-plane is represented in fig. 7.6. The poles are located on the curve $h_1$, one of the *Stokes lines* across which the asymptotic behavior of this function changes (Dingle 1973). This curve meets the real axis at $\lambda = \beta$, at an angle of 60° (fig. 7.6), and is asymptotically parallel to the imaginary axis. The lowest-order poles $\lambda_n$ are approximately given by (7.17), where $\mathrm{Im}\,\lambda_n$ grows rapidly with $n$.

The asymptotic behavior of the cylindrical functions $Z_\lambda(x)$ as $|\lambda| \to \infty$ is fairly complicated. We refer to Appendix A in Nussenzveig 1965 for a comprehensive discussion.

The purpose of the path deformations in the CAM method, as has already been mentioned in Sec. 6.2, is to concentrate dominant contributions to the integrals into the neighborhood of a small number of points in the $\lambda$ plane, that will be referred to as *critical points*. The main types of critical points are *saddle points*, that can be real or complex, and *complex poles* such as the Regge poles $\lambda_n$.

A real saddle point is also a stationary-phase point. Since the phase of the integrand in most cases can be approximated by the WKB phase, stationarity implies that the action principle (or Fermat's principle) is satisfied, so that such points will be associated with contributions from *classical paths* (or *geometric-optic rays*).

For example, for the hard-sphere scattering amplitude, obtained by substituting (7.2) into (1.14), one finds a saddle point

$$\overline{\lambda} = \beta \cos(\theta/2) \qquad (7.21)$$

The corresponding impact parameter, according to the localization principle, is $b = \overline{\lambda}/k = a \cos(\theta/2)$. Referring to fig. 7.1, we see that this is precisely the impact parameter associated with an incident ray that gets geometrically reflected in the direction $\theta$, and the saddle-point contribution is found to be equivalent to the WKB result.

To interpret Regge pole contributions, let us consider the residue at a Regge pole $\lambda_n$ of an integrand containing an angular function $P_{\lambda-\frac{1}{2}}(\cos\theta)$. Just as the decomposition $J_\lambda(kr) = [H_\lambda^{(1)}(kr) + H_\lambda^{(2)}(kr)]/2$ represents a standing radial wave function as a superposition of incoming and outgoing waves, this standing angular function can be decomposed into running angular waves by setting (Nussenzveig 1965)

$$P_{\lambda-\frac{1}{2}}(\cos\theta) = Q_{\lambda-\frac{1}{2}}^{(1)}(\cos\theta) + Q_{\lambda-\frac{1}{2}}^{(2)}(\cos\theta) \qquad (7.22)$$

in which the Legendre functions $Q_{\lambda-\frac{1}{2}}^{(1,2)}(\cos\theta)$ (Robin 1958), not too close to forward or backward directions, have the asymptotic behavior

$$Q_{\lambda-\frac{1}{2}}^{(1,2)}(\cos\theta) \approx \frac{\exp\{\mp i[\lambda\theta - (\pi/4)]\}}{(2\pi\lambda\sin\theta)^{\frac{1}{2}}} \qquad (|\lambda|\sin\theta \gg 1) \qquad (7.23)$$

where upper and lower signs correspond to superscripts 1 and 2, respectively.

Therefore, $Q_{\lambda-\frac{1}{2}}^{(1)}(\cos\theta)$ represents a clockwise-traveling angular wave and $Q_{\lambda-\frac{1}{2}}^{(2)}(\cos\theta)$ an anticlockwise-traveling one. In a Regge pole residue, they give rise to terms like $\exp(i\lambda_n\theta)$ and $\exp[i\lambda_n(2\pi-\theta)]$, the latter arising from a combination with one factor $\exp(2i\pi\lambda)$ from the Poisson sum formula, together with extra Poisson terms adding $2m\pi$ to the angle. Thus, they lead to expressions of the form (7.7): the hard-sphere Regge poles are associated with *surface wave contributions*.

We see that, as was stated in Sec. 7.2, the Regge poles yield the propagation exponents of the creeping modes that correspond to the diffracted rays of the geometrical theory of diffraction. At the same time, the diffraction coefficients $D_n$ in (7.7) are determined by the residues at the corresponding Regge poles.

The path deformations that are allowed by the asymptotic behavior of the integrands in the $\lambda$ plane differ depending on the spatial domain where the observation point lies. Typically, an allowed deformation may lead to sweeping over Regge poles, thus adding the corresponding residue contributions, and resulting in a *background integral* over a deformed path (cf. Sec. 6.2) Actually, in view of the infinite number of poles, one may have to consider a sequence of paths with increasing radii, passing halfway between consecutive Regge poles (fig. 7.6).

When the integrand has saddle points, the asymptotic behavior of the background integral is usually dominated by their contributions, which may be evaluated by steepest descent methods (Olver 1974), yielding *WKB contributions*.

In different regions, different splittings of the Poisson terms [e.g., by applying (7.22) or relationships connecting angular functions with arguments $\cos\theta$ and $-\cos\theta$], as well as different paths are generally required. The domain of applicability of a given asymptotic representation can be determined by estimating the magnitude of correction terms, and the choice of paths is guided by the location of critical points and the asymptotic behavior of the integrands.

It is at the boundaries between different regions that the most interesting diffraction effects tend to be found. In the simplest version of the CAM method, they are treated by transitional approximations, but for a smooth match with results in adjoining regions uniform approximations are usually required. We now describe the results obtained by applying the transitional version of these procedures to the hard sphere problem, referring to Nussenzveig 1965 for their derivation.

## 7.5.　Structure of the hard sphere wave function

There are six main regions to be considered, shown schematically in fig. 7.7. In this figure, the boundary of the geometrical shadow cylinder beyond the sphere is shown in dashed line; the coordinates of a point on the shadow boundary are $(r, \theta_0)$, where

$$\theta_0 \equiv \sin^{-1}(a/r) \tag{7.24}$$

(1) *Deep shadow.* This is the domain

$$r \ll a/\gamma, \quad \theta_0 - \theta \gg \gamma \tag{7.25}$$

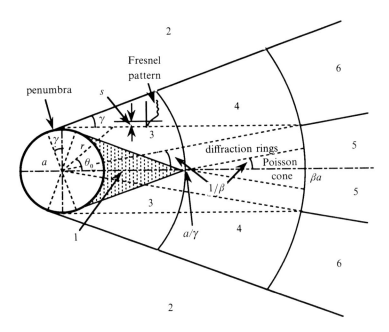

Fig. 7.7. Division into regions for a hard sphere (the angles are greatly exaggerated). (1) Deep shadow; (2) Deep illumination; (3) Fresnel region: a Fresnel pattern shows the shift $s$ of the shadow boundary; (4) Fresnel–Lommel region, showing the birth of the Poisson cone, surrounded by diffraction rings; (5) Forward diffraction peak; (6) Fock transition region.

represented by the shaded region in fig. 7.7. It is only within this region that the original Watson transformation can be applied, reducing the wave function to a rapidly convergent residue series (Sec. 6.2). The wave function is exponentially small, behaving like a superposition of surface wave contributions, associated with diffracted rays. Much weaker damping within the shadow is found behind a circular disc (Jones 1964): the shadow thrown by the sphere is much darker. Thus, classical diffraction theory, which predicts the same behavior for the sphere and for the disc, fails in this domain.

(2) *Deep illumination.* This is the domain

$$\theta - \theta_0 \gg \gamma \qquad (7.26)$$

where the wave function is dominated by the WKB approximation. It represents the superposition of the incident and geometrically reflected waves, with correction terms that become large when (7.26) is not satisfied. One also finds the continuation of the surface waves from the deep shadow, but it is exponentially small in comparison with the WKB terms, except very close to the shadow boundary. On the surface of the sphere, Kirchhoff's approximation (7.14) to the normal derivative of the wave function is accurate, except within the penumbra region that separates the shadow and illuminated hemispheres. Within the penumbra, which has an angular width of order $\gamma$ (fig. 7.7), one obtains the Fock result (7.15) as a transitional approximation.

(3) *Fresnel region.* This neighborhood of the shadow boundary is defined by

$$|\theta - \theta_0| \ll \gamma, \quad \gamma a \ll z = r\cos\theta \ll a/\gamma \tag{7.27}$$

The angular diffraction pattern in this region is found to be very similar to the Fresnel diffraction pattern of a straight edge, as illustrated in fig. 7.7. Classical diffraction theory predicts a pattern similar to that of a slit, including a contribution from the far 'edge' which here is cut off by the exponential damping in the deep shadow. Thus, classical diffraction theory fails again in this region. There are corrections to the Fresnel pattern, arising from the curvature of the sphere, that take the form of Fock-type functions. One of these corrections is a small shift of the shadow boundary, located, as is done in classical diffraction theory, by the condition that $|\psi|^2 = 1/4$ on the boundary. This shift, denoted by $s$ in fig. 7.7, is of the order of $\gamma^2 a$.

(4) *Fresnel-Lommel region.* This domain, defined by

$$\theta \lesssim \theta_0, \quad a/\gamma \ll r \ll \beta a \tag{7.28}$$

is the first one where classical diffraction theory yields a good approximation: the wave function approximately agrees with the classical Fresnel pattern of a circular disc, first derived by Lommel (Petiau 1955). There are small corrections, described by Fock-type functions, that play a significant role in smoothing the transition to neighboring regions. Along the axis, one finds a Poisson bright spot, with the same amplitude as the incident wave (Sec. 2.2). However, in contrast with the situation found for a circular disc, the spot only starts to develop at large axial distances, of the order of $a/\gamma \sim \beta^{1/3}a$. The spot grows into a bright disc corresponding to a cone with opening angle of order $1/\beta$, surrounded by diffraction rings (fig. 7.7), which gradually merges with the Airy diffraction pattern as one moves into the Fraunhofer region $r \gg \beta a$, where the last regions are located.

(5) *Diffraction peak.* This is the domain

$$0 \le \theta \lesssim 1/\beta, \quad r \gg \beta a \tag{7.29}$$

located in the Fraunhofer region (far field), where the wave function may be described in terms of the scattering amplitude $f(k, \theta)$. The main term in this region is the classical Airy pattern (7.5), dominated by the forward diffraction peak. However, there are (smaller) Fock corrections, that modify the total cross section [cf. (7.6)], the lowest one being of order $\gamma^2$, like the shift of the shadow boundary.

(6) *Fock transition region.* This domain, defined by

$$1/\beta \lesssim \theta \lesssim \gamma, \quad r \gg \beta a \tag{7.30}$$

corresponds to the transition between the diffraction peak region and the deep illuminated region $\theta \gg \gamma$, where the WKB approximation is valid and the scattering amplitude is dominated by the geometrical reflection amplitude (7.3). A transitional approximation in this domain leads to Fock-type functions, which, as usual, do not quite bridge the gap between the adjacent domains. The procedures developed to deal with this problem will be discussed in the next chapter.

More mathematically oriented treatments and extensions of these results on the behavior of the wave function at finite distances may be found in Ludwig 1969 and Melrose & Taylor 1986.

# Diffraction as tunneling

We now show how CAM theory deals with the first of the semiclassical critical effects, forward diffractive scattering. The impenetrable sphere model is ideal for this purpose, because the only relevant physical processes that take place in it are reflection from the surface and forward diffractive scattering.

The difficulties are connected with the transition domain between the forward diffraction peak and the wide-angle geometrical reflection region, where the WKB approximation holds (Sec. 7.5). Within the forward peak, classical diffraction is dominant, but this domain of small diffraction angles is only sensitive to the bulk blocking effect of the scatterer (Sec. 2.2). To probe the dynamics of diffraction one must go to larger diffraction angles, contained within this difficult transition domain.

Fock's theory of diffraction is not powerful enough to bridge the gap, as will be seen in several examples. For this purpose, one needs a *uniform asymptotic approximation*. Such an approximation was developed by Berry (1969) for forward diffractive scattering by potentials with long-range tails, but near-forward scattering in this case arises from large impact parameters, a physical mechanism completely different from diffraction by the edge of a curved surface.

The crucial physical concept is that of the *effective potential*. It will be seen to lead to a reinterpretation of Fock's theory, clarifying several points that remained obscure in it (Sec. 7.3). The uniform CAM results, besides being extremely accurate and having a very broad range of validity that extends even to the long-wavelength region, lead to a new physical picture of the dynamics of diffraction. It will be seen that *tunneling* plays an important role in this process; this is also true for all other critical effects in semiclassical scattering, as will appear in succeeding chapters.

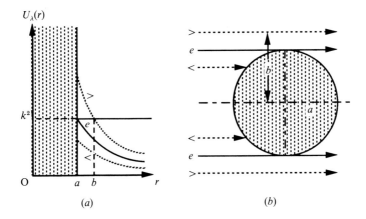

Fig. 8.1. *(a)* Effective potential $U_\lambda(r)$ for three different values of the angular momentum: edge *(e)*, above-edge *(>)*, with turning point *b*, and below-edge *(<)*. *(b)* Corresponding incident rays.

## 8.1.  Effective potential and edge domain

The hard sphere effective potential $U_\lambda(r)$ for radial motion with angular momentum $\lambda$ is the sum of the hard core and the centrifugal 'potential' (in the Langer sense; cf. Sec. 1.2) $\lambda^2/r^2$. We take units such that $\hbar = 2m = 1$, so that the energy is $k^2$.

This effective potential is schematically represented in fig. 8.1*(a)* for three different values of the angular momentum, associated with three different impact parameters $b(\lambda) = \lambda/k$ by the localization principle. The corresponding incident rays are shown in Fig. 8.1*(b)*.

The *edge angular momentum*

$$\lambda_e \equiv \beta \tag{8.1}$$

is associated with an impact parameter $a$, i.e., with incident *edge rays*, tangential to the sphere. As shown in fig. 8.1*(a)*, the corresponding energy level $k^2$ sits right at the top of the centrifugal barrier.

A *below-edge* angular momentum $\lambda < \lambda_e$ represents a lower centrifugal barrier, so that the incident ray hits the sphere and the radial turning point remains at $r = a$. For an *above-edge* angular momentum $\lambda > \lambda_e$, the energy is below the barrier top, so that the radial turning point, where the centrifugal barrier is met, coincides with the impact parameter

$b > a$, representing an incident ray that passes outside of the sphere [fig. 8.1(b)]. It is essential for the discussion that follows to keep these associations well in mind.

As should be clear from fig. 8.1(a), in the above-edge situation, one must take into account the effects of *tunneling through the centrifugal barrier*: if the energy level is close enough to the barrier top, it should still be possible to penetrate all the way to the surface with appreciable amplitude. The WKB barrier penetration factor from $r = b$ to $r = a$ is given by [Berry & Mount 1972; cf. (1.18)]

$$\exp\left[-\int_{\beta}^{\lambda}\left(\frac{\lambda^2}{x^2}-1\right)^{1/2}dx\right] \sim \exp\left\{-\frac{1}{3}[\gamma(\lambda-\beta)]^{3/2}\right\} \qquad (8.2)$$

where $\gamma = (2/\beta)^{1/3}$ is the penumbra scale width, and the approximation applies when $\lambda - \beta \ll \beta$.

It follows that (Nussenzveig 1969a) the barrier transmissivity is appreciable when

$$\beta \le \lambda \le \beta\left(1 + c_+\gamma^2\right) \qquad (8.3)$$

where $c_+$ is a positive numerical constant of order unity. Similarly, in the below-edge situation, where the incident wave is reflected from the hard core, the reflection amplitude is affected by the presence of the centrifugal barrier when its top lies close enough to the energy level [fig. 8.1(a)]. The WKB criterion for this to happen is

$$\beta\left(1 - c_-\gamma^2\right) \le \lambda \le \beta \qquad (8.4)$$

where the positive numerical constant $c_-$ is also of order unity.

Conditions (8.3) and (8.4) define an anomalous neighborhood of the edge angular momentum that will be called the *edge strip*: (8.3) is the *above-edge strip* and (8.4) is the *below-edge strip*. The same designations will be applied to the corresponding ranges of impact parameters. It is readily seen that the intersection of the below-edge strip with the surface of the sphere corresponds to the penumbra region in fig. 7.7.

The edge strip plays an essential role in the new dynamical diffraction effects found in CAM theory. The importance of edge phenomena was emphasized by van de Hulst (1957).

## 8.2. Reinterpretation of the Fock theory

We saw in Sec. 7.3 how, for free wave propagation along one direction in the semiclassical regime, the wave amplitude along a propagating wavefront obeys the Schrödinger-like free-particle equation (7.12) in transverse directions.

Let us now consider, for simplicity, the scattering of a plane wave by an impenetrable circular cylinder (equivalent to a two-dimensional problem). With reference to fig. 7.5, we introduce, following Fock and Wainstein (Fock 1965), the *ray coordinates* $(\xi,\eta)$ of the observation point B in that figure, by

$$\eta = \overset{\frown}{T_2 T_2'} = a\varepsilon , \quad \xi = \eta + \overline{T_2'B} \tag{8.5}$$

so that $\eta$ is the arc length described by the corresponding diffracted ray on the surface and $\xi$ is the total diffracted ray path from the edge to the observation point. Since $\xi = $ constant defines a diffracted wavefront ABCD (fig. 7.5) and $\eta = $ constant describes the straight rays tangentially leaving the surface, this is a set of orthogonal coordinates.

We further introduce the scaled dimensionless variables

$$x \equiv \frac{\xi/a}{\gamma} , \quad y \equiv \left[ \frac{(\xi - \eta)/a}{\gamma} \right]^2 \tag{8.6}$$

where $\gamma$ is the penumbra scale width. Then, $\partial/\partial x$ refers to evolution along a diffracted ray (the analogue of time evolution in the picture of Sec. 7.3) and $y$ plays the role of a transverse coordinate: it vanishes on the surface and it increases as we move away from it.

It may then be shown that, in a neighborhood of the edge $T_2$ (fig. 7.5), the reduced wave equation leads to an amplitude equation of the form (Nussenzveig 1988)

$$i \,\partial A/\partial x = \left( -\partial^2/\partial y^2 - y \right) A \tag{8.7}$$

Comparing this with (7.12), we see that one now gets the analogue of the time-dependent Schrödinger equation for motion in a *linear potential barrier*. Since one is still in free space, this is an *inertial* term, arising from the effects of the edge curvature described through the ray coordinates.

Going back to fig. 8.1(a), we can immediately interpret these results: near the curved edge, one is just probing the top portion of the centrifugal barrier, which may be approximated by a linear potential. Thus, *it is indeed the Schrödinger equation, with an effective potential arising from the curvature, that describes the dynamics of diffraction*: the proposed analogy with diffusion is misleading.

The 'stationary' solutions of (8.7), as is well known, are Airy functions (Abramowitz & Stegun 1965). The general solution can be built up as a wave packet of stationary solutions. For the scattering problem, the wave packet expansion coefficients are determined by requiring the solution to satisfy the boundary conditions at the surface, as well as the excitation conditions associated with the incident plane wave.

The result is found to be given by the Fock approximation of Sec. 7.3. This explains the role of the Airy functions and the form in which they appear in the Fock integrals. At the same time, we see that the limitations of Fock's theory arise from employing just a linear approximation to the effective inertial potential. In order to extend the solution far enough outside of the penumbra so as to merge smoothly with the WKB approximation in the geometrical reflection region, a more powerful approach is required.

## 8.3.  Outer and inner representations

From now on we shall not discuss the wave function at finite distances, restricting our discussion to the scattering amplitude. As was explained in Sec. 7.4, one needs different CAM representations in angular domains that extend all the way to the backward direction from those applicable all the way to the forward direction; at intermediate angles these alternative representations are equivalent and may be transformed into each other.

We refer to the backward-extensible representation, in terms of angular functions of the form $P_{\lambda-\frac{1}{2}}(-\cos\theta)$, as *outer representation*, and to the forward-extensible one, expressed in terms of $P_{\lambda-\frac{1}{2}}(\cos\theta)$, as *inner representation*. We quote here only the form of these representations (for their derivation and additional details, see Nussenzveig 1988).

We employ the dimensionless scattering amplitude $F(\beta,\theta) \equiv f(k,\theta)/a$. The outer representation, valid for $0 < \theta \le \pi$, but to be employed for $\gamma \ll \theta \le \pi$, is of the form

$$F(\beta,\theta) = F_r(\beta,\theta) + F_s(\beta,\theta) \qquad (8.8)$$

where

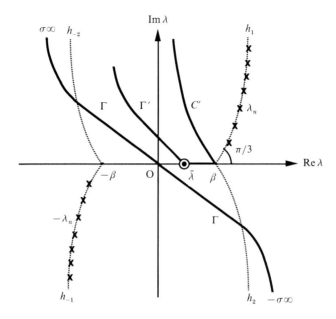

Fig. 8.2.  Paths of integration in the $\lambda$ plane. ⊙ Saddle point $\bar{\lambda}$ ; **x** Regge poles.

$$F_r(\beta,\theta)=(i/\beta)\int_{\sigma\infty}^{0}S(\lambda,\beta)P_{\lambda-\frac{1}{2}}(-\cos\theta)\tan(\pi\lambda)e^{-i\pi\lambda}\lambda\,d\lambda$$

$$=-(i/\beta)\int_{\sigma\infty}^{-\sigma\infty}S(\lambda,\beta)Q_{\lambda-\frac{1}{2}}^{(1)}(\cos\theta)\lambda\,d\lambda \tag{8.9}$$

is the *reflection amplitude*.

The path of integration in the second line of (8.9) is the path $\Gamma$ in fig. 8.2, which is symmetric about the origin, allowing it to be transformed into the first line, that is explicitly regular at $\theta = \pi$. The $S$-function is given by (7.20).

The other term in (8.8) is the *surface-wave amplitude*

$$F_s(\beta,\theta)=-\frac{2\pi i}{\beta}\sum_{n=1}^{\infty}\lambda_n r_n\frac{\exp(i\pi\lambda_n)}{1+\exp(2i\pi\lambda_n)}P_{\lambda_n-\frac{1}{2}}(-\cos\theta) \tag{8.10}$$

where the sum is extended over all Regge poles in the first quadrant of the $\lambda$ plane and $r_n$ is the residue of $S$ at $\lambda_n$.

The inner representation, valid for $0\le\theta<\pi$, is employed in the domain complementary to that of the outer one. It is of the form

$$F(\beta,\theta)=F_b(\beta,\theta)+F_{s'}(\beta,\theta)+F_{e+}(\beta,\theta)+F_{e-}(\beta,\theta) \tag{8.11}$$

Here, $F_b(\beta, \theta)$ is the *blocking amplitude*, given by

$$F_b(\beta, \theta) = \frac{i}{\beta}\int_0^\beta P_{\lambda-\frac{1}{2}}(\cos\theta)\,\lambda\,d\lambda$$

$$-\frac{2i}{\beta}\int_0^{i\infty}\frac{\exp(2i\pi\lambda)}{1+\exp(2i\pi\lambda)}P_{\lambda-\frac{1}{2}}(\cos\theta)\,\lambda\,d\lambda \qquad (8.12)$$

which does not contain the *S*-function at all.

The second term of (8.11), the *surface-wave contribution*, differs from (8.10) only by the substitution

$$P_{\lambda_n-\frac{1}{2}}(-\cos\theta) \rightarrow -i\exp(i\pi\lambda_n)P_{\lambda_n-\frac{1}{2}}(\cos\theta)$$

The last two terms of (8.11), together, represent the *edge amplitude*.

The first contribution to the edge amplitude is the *above-edge amplitude*, given by

$$F_{e+}(\beta, \theta) = (i/\beta)\int_\beta^\infty[1 - S(\lambda, \beta)]P_{\lambda-\frac{1}{2}}(\cos\theta)\,\lambda\,d\lambda \qquad (8.13)$$

Finally, the last term of (8.11), the *below-edge amplitude*, is given by

$$F_{e-}(\beta, \theta) = (-i/\beta)\int_{i\infty}^\beta S(\lambda, \beta)P_{\lambda-\frac{1}{2}}(\cos\theta)\,\lambda\,d\lambda \qquad (8.14)$$

taken over a path of integration, such as $C'$ in fig. 8.2, that stays to the left of the Regge poles.

All these representations are *exact*. The next step is the derivation of asymptotic approximations to the various terms. This also brings out their physical interpretation, justifying the names that have been given to them. Referring to Nussenzveig 1988 for the derivations, we now discuss the results obtained.

## 8.4  The CAM approximation

The *outer CAM approximation* is obtained as an asymptotic approximation to the outer representation (8.8). The *uniform CAM approximation* is an asymptotic approximation to the inner representation (8.11). The *CAM approximation* is obtained by matching these two approximations at a suitably chosen intermediate scattering angle where both of them can be applied.

## Outer CAM approximation

(i) *Reflection amplitude.* The reflection amplitude (8.9) is dominated by a saddle-point contribution arising from the geometrical reflection saddle point (7.21). The steepest-descent evaluation leads to the *WKB expansion*

$$F_r(\beta,\theta) = -\frac{1}{2}\left\{1+\frac{i}{2\beta\sin^3\frac{1}{2}\theta}+\frac{2+3\cos^2\frac{1}{2}\theta}{\left(2\beta\sin^3\frac{1}{2}\theta\right)^2}\right.$$

$$\left.+O\left[\left(2\beta\sin^3\frac{\theta}{2}\right)^{-3}\right]\right\}\exp\left(-2i\beta\sin\frac{\theta}{2}\right) \qquad (8.15)$$

the first two terms of which were already given in (7.3). For small $\theta$, this is an expansion in inverse powers of $(\theta/\gamma)^3$, so that the expansion breaks down in the penumbra region $\theta \lesssim \gamma$.

(ii) *Surface-wave amplitude.* Only the lowest-order Regge poles, located near $\lambda = \beta$ (fig. 8.2), give a significant contribution to (8.10). For large $\beta$, the poles and their corresponding residues are given by the asymptotic expansions

$$\lambda_n = \beta + e^{i\pi/3}x_n\gamma^{-1} + e^{2i\pi/3}\frac{x_n^2}{60}\gamma + O(\gamma^3) \qquad (8.16)$$

$$r_n = \frac{e^{-i\pi/6}}{2\pi a_n'^2\gamma}\left[1+\frac{e^{i\pi/3}}{30}x_n\gamma^2 + O(\gamma^4)\right] \qquad (8.17)$$

where $x_n$ is the $n$th zero of Ai$(-x)$ [cf. (7.17)] and $a_n' \equiv$ Ai$'(-x_n)$.

In particular, for $\pi - \theta \gg \beta^{-1}$, (8.10) is dominated by

$$F_s(\beta,\theta) \approx \frac{e^{-5i\pi/12}}{2}\left(\frac{\gamma}{\pi\sin\theta}\right)^{\frac{1}{2}}e^{i\beta\theta}\sum_n a_n'^{-2}\exp\left[-\frac{\left(\sqrt{3}-i\right)x_n\theta}{2\gamma}\right] \qquad (8.18)$$

which, by comparison with the first term of (7.7), yields the diffraction coefficients $D_n$. The only domain where the surface-wave contribution may become appreciable is close to the penumbra, i.e., at the boundary of the domain of applicability of the outer CAM approximation.

### Uniform CAM approximation

(i) *Surface-wave contribution.* The second term of (8.11) gives contributions similar to (8.18), except by the replacement of $\theta$ by $2\pi - \theta$. This represents surface waves that have already taken at least half a turn around the sphere, so that their effect is negligible, except possibly at low values of $\beta$.

(ii) *Blocking amplitude.* Employing the Szegö–Olver uniform asymptotic expansion of the Legendre function (Olver 1974)

$$P_{\lambda-\frac{1}{2}}(\cos\theta) \approx \mathscr{P}(\theta,\lambda) \equiv \left(\frac{\theta}{\sin\theta}\right)^{\frac{1}{2}} \left\{\left[1+O\left(\lambda^{-2}\right)\right]J_0(\lambda\theta)\right.$$
$$\left. -\tfrac{1}{8}\left(\theta^{-1}-\cot\theta\right)J_1(\lambda\theta)/\lambda + O\left(\lambda^{-3}\right)\right\} \quad (8.19)$$

one obtains from (8.12)

$$F_b(\beta,\theta) = \mathrm{i}\left(\frac{\theta}{\sin\theta}\right)^{\frac{1}{2}}\left\{\frac{J_1(\beta\theta)}{\theta}\right.$$
$$\left. -\tfrac{1}{8}\left(\theta^{-1}-\cot\theta\right)\frac{\left[1-J_0(\beta\theta)\right]}{\beta\theta} + O\left(\beta^{-1}\right)\right\} \quad (8.20)$$

which is the uniform expression for the *classical Airy pattern* (7.5). Within the forward peak, the first term within the curly bracket, which is dominant, differs from (7.5) only by the replacement of $\sin\theta$ by $\theta$. As was already observed, (8.12) is the CAM expression of Fresnel's *blocking effect*, which depends only on the scatterer size: the $S$-function does not enter at all.

The new diffraction effects are contained in the last two terms of (8.11), the edge contributions:

(iii) *Above-edge amplitude.* By employing (8.19) as well as Olver's uniform asymptotic expansions of the cylindrical functions (Olver 1974), one obtains from (8.13)

$$F_{e+}(\beta,\theta) = -\int_0^\infty \mathscr{H}(x,\varphi)\,\mathscr{P}(\theta,\beta\cosh\varphi)(\cosh\varphi/\varphi)\sqrt{x}\,\mathrm{d}x \quad (8.21)$$

where $\mathscr{P}(\theta, \lambda)$ is defined by (8.19),

$$\mathscr{H}(x, \varphi) \equiv \frac{e^{-i\pi/6}\mathrm{Ai}(x) - \sigma(x, \varphi)\mathrm{Ai}'(x)}{\mathrm{Ai}(e^{2i\pi/3}x) + e^{-i\pi/6}\sigma(x, \varphi)\mathrm{Ai}'(e^{2i\pi/3}x)} \qquad (8.22)$$

with

$$\sigma(x, \varphi) \equiv \frac{5e^{-i\pi/6}}{24\beta\sqrt{x}\sinh\varphi}\left[\frac{1}{3(\varphi\coth\varphi - 1)} - \coth^2\varphi + \frac{3}{5}\right] \qquad (8.23)$$

and

$$x \equiv \left[\tfrac{3}{2}\beta(\varphi\cosh\varphi - \sinh\varphi)\right]^{2/3} \qquad (8.24)$$

which can be inverted (Nussenzveig 1988) to yield $\varphi$ as a power series in $\gamma x^{1/2}$. The appearance of both $\mathrm{Ai}(x)$ and $\mathrm{Ai}'(x)$ in (8.22) is typical of uniform asymptotic expansions, as is that of $J_0$ and $J_1 = -J_0'$ in (8.19).

The variable $x$ in (8.21) is equivalent to $\gamma(\lambda - \beta) \equiv \gamma k(b - a)$, where $b$ is the impact parameter associated with $\lambda$, and $|x| \lesssim 1$ defines the edge strip (Sec. 8.1). For $x \gg 1$, (8.22) is dominated by

$$e^{i\pi/3}\mathscr{H}(x, \varphi) \approx \exp\left(-\tfrac{4}{3}x^{3/2}\right) = \left[\exp\left(-\int_a^b\sqrt{\frac{\lambda^2}{r^2} - k^2}\right)\right]^2 \qquad (8.25)$$

where the impact parameter $b = \lambda/k$ is the radial turning point (fig. 8.1). The expression within square brackets is the WKB barrier penetration factor (8.2), and (8.22) is just the uniform version of this factor. Thus, *the above-edge amplitude represents the effect of uniform tunneling of the outside rays through the centrifugal barrier to the surface (with which they interact) and back. It is a pure tunneling amplitude.*

(iv) *Below-edge amplitude.* This amplitude, given by an asymptotic approximation to (8.14), describes the diffraction effects associated with the anomalous reflection, at near-glancing incidence, of rays in the edge domain, in view of the proximity to the top of the centrifugal barrier in fig. 8.1(*a*); equivalently, it contains the *effects of surface curvature on reflection*. This has some analogy to Young's ideas on diffraction as a 'peculiar kind of reflection'.

At $\theta = 0$, the WKB saddle point (7.21) would be at $\lambda = \beta$, corresponding to the edge ray. This is *not* a saddle point of (8.14). However, the path $C'$ in (8.14) and fig. 8.2 is analogous to a steepest-descent path; the fast decay of the integrand may be associated with tunneling for complex angles of incidence. As $\theta$ increases, one finds that, up to $\theta \sim \gamma$, $C'$ remains a good path. However, for $\theta > \gamma$, it is better to switch to the path $\Gamma'$ (fig. 8.2), which is the upper half of the steepest-descent path through the saddle point $\bar{\lambda}$, followed by the interval of the real axis between $\bar{\lambda}$ and $\beta$, which is a portion of a stationary-phase path. In this way, we follow the 'birth' of the reflected wave in the WKB region: one half of it arises from the steepest-descent path and the other half from the stationary-phase path.

## 8.5  The Fock approximation

One may go over from the uniform CAM approximation to the Fock approximation by employing power series expansions of the uniform results for the edge contributions (valid for large enough $\beta$ and small enough $\theta$) and by replacing the uniform asymptotic expansions of the radial and angular functions by transitional asymptotic expansions.

The results can be expressed in terms of *generalized Fock functions*, defined by (Nussenzveig 1969a)

$$F_{m,n}(\beta,\theta) \equiv \frac{e^{i\pi/6}}{2\pi} \int_C \frac{z^m J_n[\beta\theta + (\theta/\gamma)z]}{\mathrm{Ai}^2(e^{2i\pi/3}z)} dz \qquad (8.26)$$

where $J_n$ is the Bessel function, the path $C$ runs from $e^{2i\pi/3}\infty$ to 0 and from 0 to $\infty$, and $m$ and $n$ are integers.

The result (Nussenzveig 1988) is

$$F_{e+}(\beta,\theta) + F_{e-}(\beta,\theta) \approx i\left(\frac{\theta}{\sin\theta}\right)^{1/2}\left[-\frac{J_1(\beta\theta)}{\theta} + \frac{F_{0,1}(\beta,\theta)}{\theta}\right.$$
$$\left. + \frac{\gamma^2}{2\theta}F_{1,1}(\beta,\theta) + \frac{\gamma}{60}F_{2,0}(\beta,\theta)\right] \qquad (8.27)$$

where it is assumed that $\theta \lesssim \gamma \ll 1$. For $\theta \ll \gamma$, this can be expanded into powers of $\theta/\gamma$:

$$F_{e+}(\beta,\theta) + F_{e-}(\beta,\theta) \approx i\left(\frac{\theta}{\sin\theta}\right)^{\frac{1}{2}} \left\{\left[\frac{M_0}{\gamma} + \frac{8}{15}M_1\gamma\right.\right.$$

$$\left.\left. -\frac{M_2}{2\gamma}\left(\frac{\theta}{\gamma}\right)^2 + \cdots\right]J_0(\beta\theta) - \left(\frac{M_1}{\gamma} + \frac{3}{10}M_2\gamma + \cdots\right)\frac{\theta}{\gamma}J_1(\beta\theta) + \cdots\right\} \quad (8.28)$$

where

$$M_n \equiv e^{i\pi/3} \int_0^\infty \frac{x^n \text{Ai}(x)}{\text{Ai}\left(e^{2i\pi/3}x\right)}\,dx + e^{i(2n+1)\pi/3}\int_0^\infty \frac{x^n \text{Ai}(x)}{\text{Ai}\left(e^{-2i\pi/3}x\right)}\,dx \quad (8.29)$$

These *Fock coefficients* were introduced by Wu (1956). The first few of them are (Lisle *et al.* 1985)

$$M_0 = 1.255\,124\,56\,e^{i\pi/3}\,, \quad M_1 = 0.532\,290\,7\,e^{2i\pi/3}\,, \quad M_2 = 0.067\,717 \quad (8.30)$$

Substituting these results into (8.11) and employing the optical theorem (7.6), one obtains the Fock approximation to the total cross section:

$$\sigma/\left(\pi a^2\right) = 2 + 2\,\text{Re}\left[M_0\gamma^2 + \frac{8}{15}M_1\gamma^4 + \left(\frac{4}{175}M_2 + \frac{23}{1680}\right)\gamma^6 + O\left(\gamma^8\right)\right] \quad (8.31)$$

which gives the edge correction to the classical 'extinction paradox' result of Sec. 7.1 as a power series in $\gamma^2$.

## 8.6   Numerical comparisons

Numerical summation of the partial-wave series for large size parameters is a major task, full of pitfalls and making considerable demands on computer time (which grows roughly linearly with the size parameter), even with techniques refined through decades of programming advances (Wiscombe 1979, 1980, Bohren & Huffman 1983). The convergence pattern is complicated and oscillatory, and convergence only comes about in the neighborhood of the cutoff value (5.10). The typical accuracy achieved is of the order of 1 ppm (part per million).

For the CAM approximation, only a few terms have to be evaluated. Even for size parameters of order unity, at most four terms need to be kept in the residue series (8.10). The only nontrivial numerical

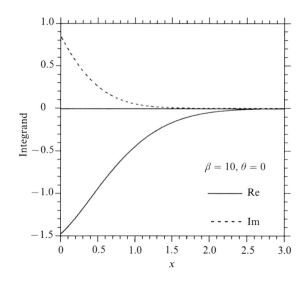

Fig. 8.3. Behavior of the above-edge integrand at $\theta = 0$ (after Nussenzveig & Wiscombe 1991).

work is the evaluation of the edge amplitudes. This involves one definite integral with a Fresnel-like oscillatory integrand, but at most a few oscillations within the range of integration for the angular domain where it is needed (the integral from $\overline{\lambda}$ to $\beta$ in fig. 8.2), and a couple of infinite integrals with smooth and rapidly decreasing integrands (faster than exponential decrease). The procedure is size-independent and numerically stable.

The real and imaginary parts of the above-edge integrand that appears in (8.21), for $\theta = 0$, are plotted (Nussenzveig & Wiscombe 1991) in fig. 8.3. This essentially represents the uniform tunneling amplitude (8.22), which decreases very fast with $x = \gamma(\lambda - \beta) \equiv \gamma k(b - a)$, as indicated in (8.25). We see that *the tunneling range of impact parameters beyond the edge that contributes significantly is*

$$b - a \sim \left(\lambda_0^2 a\right)^{\frac{1}{3}} \tag{8.32}$$

where $\lambda_0 \equiv 2\pi/k$ is the wavelength.

Numerical tests comparing the CAM and Fock approximations with 'exact' partial-wave results have been performed for several models of impenetrable sphere scattering, in quantum mechanics, acoustics and electromagnetic scattering, with comparable results in all cases. We illustrate them by a couple of examples (Nussenzveig & Wiscombe 1987) pertaining to acoustic scattering by a perfectly rigid sphere, which differs

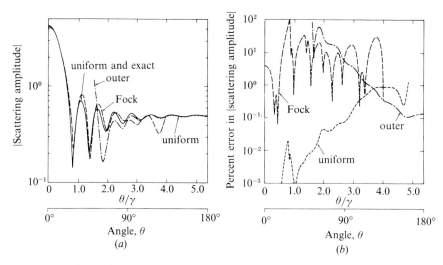

Fig. 8.4. *(a)* $|F(\beta,\theta)|$ for acoustic scattering by a rigid sphere, $\beta = 10$: exact vs. uniform, outer and Fock approximations. *(b)* Magnitude of the percent errors of the various approximations (after Nussenzveig & Wiscombe 1987).

from the quantum hard sphere by having a Neumann rather than Dirichlet boundary condition (vanishing normal derivative at the surface).

In fig. 8.4*(a)*, for $\beta = 10$, we compare the 'exact' partial-wave results for $|F(\beta,\theta)|$ with the uniform and outer CAM approximations, as well as with the Fock approximation. The angular distribution already shows a sharp forward peak and several diffraction oscillations. In the outer approximation, these oscillations arise from interference between the WKB (geometrical reflection) and surface-wave contributions, so that surface-wave effects are quite visible here.

The uniform approximation can only be distinguished from the exact curve for $\theta \gtrsim 4\gamma$, while for the outer approximation the departure becomes visible only for $\theta \lesssim 3.5\gamma$. The CAM approximation, obtained by matching them, gives a visually perfect fit to the exact curve. For the Fock approximation, however, departures are visible even within the forward diffraction peak, becoming very large for $\theta \gtrsim 2\gamma$, so that it cannot provide a smooth match with the geometrical reflection region.

The magnitude of the percent errors associated with the various approximations is plotted in fig. 8.4*(b)*. One should switch between the outer and uniform approximations at the crossover point between their respective error curves, that occurs at $\theta \approx 3.7\gamma \approx 133°$, where one finds the maximum relative error of the CAM approximation, about 0.8% here.

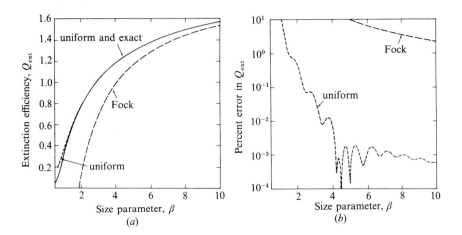

Fig. 8.5. *(a)* $\sigma(\beta)/\pi a^2$ for acoustic scattering by a rigid sphere, $1 \le \beta \le 10$: exact vs uniform and Fock approximations. *(b)* Magnitude of the percent errors of the various approximations (after Nussenzveig & Wiscombe 1987).

However, typical CAM errors are far smaller: about 0.001% within the main peak, 0.01% around the first minimum and less than 0.1% over the whole forward hemisphere, where about 70% of the total cross section is concentrated. Typical Fock approximation errors are three to four orders of magnitude higher. Very similar results are found for the phase of the scattering amplitude.

Corresponding results for the total cross section (divided by the geometrical one) are plotted in fig. 8.5 for $1 \le \beta \le 10$. The departure of the uniform CAM curve from the exact one only becomes noticeable close to $\beta = 1$, whereas the Fock approximation is quite poor in this whole range and even yields negative cross sections for low values of $\beta$. The percent error of the CAM approximation [fig. 8.5(*b*)] is about 10% at $\beta = 1$, < 1% for $\beta > 1.5$, < 0.1% for $\beta > 3$, < 0.01% for $\beta > 4$, and it has fallen to a few ppm at $\beta = 10$. By contrast, the percent error of the Fock approximation is three to four orders of magnitude larger, falling below 10% only for $\beta > 5$.

We see that *the CAM approximation bridges the gap between short-wavelength and long-wavelength scattering*, remaining applicable all the way down to $\beta \sim 1$. Indeed, at $\beta = 1$, it yields an error for the angular distribution of less than 0.01% at small $\theta$, climbing only to a few percent at larger angles (Nussenzveig & Wiscombe 1987).

On the other hand, for $\beta = 50$, one finds, for a quantum hard sphere

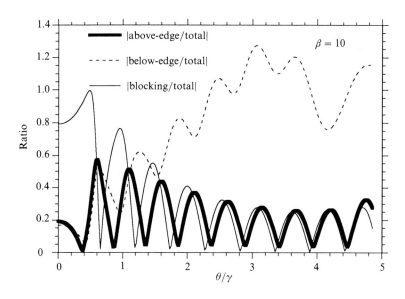

Fig. 8.6.  Relative contributions from blocking, above-edge and below-edge amplitudes to $|F(\beta,\theta)|$ for a quantum hard sphere; $\beta = 10$ (after Nussenzveig & Wiscombe 1991).

(Nussenzveig & Wiscombe 1991), that the CAM error remains below the 'noise level' of 1 ppm for $\theta < \gamma$, where most of the cross section is concentrated, rising only to a maximum of less than 0.05% at intermediate angles. For $\beta > 100$, the accuracy of the CAM approximation is better than 1 ppm, i.e., it becomes, for all practical purposes, more accurate than the 'exact' partial-wave solution!

### 8.7  Diffraction as a tunneling effect

The relative contributions from the various terms of (8.11) to $|F(\beta,\theta)|$ for a quantum hard sphere with $\beta = 10$ (the surface-wave contribution is negligible here) are plotted in fig. 8.6 (Nussenzveig & Wiscombe 1991). At $\theta = 0$, the blocking amplitude, which is the uniform representation of the Airy pattern of classical diffraction theory, is dominant, with edge corrections of order $\gamma^2$ times smaller [cf. (8.31)], the same ratio as that between the width of the edge strip (8.3)–(8.4) and the radius of the scatterer.

The dominance of blocking persists within the forward peak, but the edge terms are important near the first few minima of the Airy pattern, filling them in and decreasing the contrast of the diffraction rings;

peak locations are also displaced. Since $J_1(\beta\theta)/\theta$ decays like $\beta^{-1/2}\theta^{-3/2}$ for $\theta \gg \beta^{-1}$, the $O(\beta)$ peak enhancement at $\theta = 0$ disappears when $\theta$ becomes of order $\gamma$. Indeed, fig. 8.6 shows that for $\theta/\gamma \sim 1$ the above-edge contribution, the below-edge contribution and the contribution from classical diffraction (blocking) are all of the same order.

*Throughout the penumbra region of large diffraction angles, $\theta \gtrsim \gamma$, the above-edge (tunneling) contribution remains of the same order as that from classical diffraction (blocking).* The below-edge contribution, in this angular range, is associated with *anomalous reflection,* as was mentioned at the end of Sec. 8.4. This contribution becomes gradually more important as $\theta$ increases, until it finally merges with 'regular' geometrical reflection, described by the WKB amplitude, for $\theta \gg \gamma$.

Let us sum up the main results obtained in the CAM treatment of near-forward diffractive scattering.

(i)   The uniform CAM approximation accounts for diffraction in the penumbra region at large angles, holding all the way from $\theta = 0$ to $\theta \gg \gamma$, where it merges with the outer CAM approximation. In contrast, the Fock approximation breaks down beyond $\theta \sim \gamma$, so that it does not solve the connection problem.

(ii)  The CAM approximation is extremely accurate, giving visually perfect fits to the exact results even below $\beta = 10$. In comparison with Fock's theory, it improves the accuracy typically by several orders of magnitude.

(iii) The approximation bridges the gap between short- and long-wavelength scattering, remaining accurate all the way down to $\beta \sim 1$, both for the magnitude and for the phase of the scattering amplitude.

(iv)  In contrast with the partial-wave series, the CAM approximation is numerically stable and has a size-independent computing time. For $\beta > 100$, its accuracy already exceeds the typical 1 ppm accuracy of numerical partial-wave summations.

(v)   The CAM theory leads to a new physical picture of diffraction (Nussenzveig 1989a, 1989b). Fresnel's blocking effect, which is the basis of classical diffraction theory and which led to the quantum picture of diffraction as 'shadow scattering', is the prevailing mechanism only for small diffraction angles, within the forward peak. What happens at larger diffraction angles is strongly affected by the dynamics of diffraction, associated with the surface curvature.

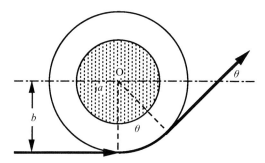

Fig. 8.7. Pseudoclassical above-edge tunneling path.

The curvature has one trivial effect: it spreads the range of angles of incidence. This is of course included in the WKB approximation, but the effect of surface curvature on reflection is neglected: as usual in geometrical optics, one substitutes the surface by its tangent plane at the point of incidence to evaluate the reflection amplitude.

The dynamical effects of curvature are manifested through the centrifugal term in the effective potential. Here one employs to full advantage the intermediate nature of the Poisson representation, poised midway between classical and quantum pictures (Sec. 6.1), to draw on both, applying classical insight to quantum problems, as well as concepts most familiar in quantum theory to classical problems. This is exemplified by the appearance of Schrödinger's equation (improperly interpreted in terms of diffusion) in Fock's theory.

The new diffraction mechanisms arising from the curvature are *tunneling* and *anomalous reflection*, both taking place within the edge strip (8.3)–(8.4). As one penetrates within the penumbra beyond $\theta \sim \gamma$, their contribution becomes comparable with that from blocking, and anomalous reflection, gradually transformed into geometrical reflection, becomes increasingly dominant.

To interpret the above-edge tunneling contributions, we extend Berry and Mount's picture of pseudoclassical paths (Sec. 6.1) to include complex paths associated with barrier penetration. A pseudoclassical above-edge tunneling path is illustrated in fig. 8.7. As the ray 'coasts' along the caustic sphere with constant angular momentum $\lambda = kb$ associated with its impact parameter, it interacts with the surface by tunneling through the centrifugal barrier, finally emerging in the direction $\theta$. This is an extended version of a diffracted ray.

The above-edge amplitude (8.21) is a superposition of such contributions, weighted by the uniform tunneling penetration amplitude, which plays the role of a form factor, plotted in fig. 8.3. The same mechanism leads to the generation of surface waves, which are therefore another manifestation of tunneling. In classical language, tunneling is described in terms of *evanescent waves*.

The new conceptual picture of edge diffraction should extend to more general smooth curved surfaces. Analogously to Sec. 8.2, a local approximation to the reduced wave equation in a suitable set of curvilinear coordinates gives rise to *inertial barriers* that generalize the centrifugal one. The tunneling of near-edge paths through such barriers should contribute significantly to large-angle diffraction around the edge. There is some similarity with the general relativistic picture of the gravitational deflection of light, though this deflection is usually treated at the level of geometrical optics; diffraction and tunneling are wave effects that take place at a much finer scale.

What is the relationship with previous ideas about diffraction? In contrast with the view that it is a purely *local* effect due to edge rays, valid in Young's and Fresnel's theories as well as in Keller's geometrical theory of diffraction, we see that tunneling gives rise to an *effectively nonlocal* (though short-ranged) interaction, with a range given by (8.32), that represents a kind of weighted geometric mean between the wavelength and the local radius of curvature. Of course, the ideas of anomalous edge reflection, blocking and diffracted rays also play a role in the overall effect.

Perhaps the most remarkable anticipation of the new physical picture of diffraction was provided by Newton in the first query of his 'Opticks' (see the quotation at the beginning of this chapter)!

# 9

# The Debye expansion

*The first mirrored the next, as though it were*
*Rainbow from rainbow*
(Dante, *Paradiso* XXXIII, 118)

We now begin to apply the CAM method to Mie scattering. The penetration of the field and the resulting interactions within the scatterer yield a rich structure, where all semiclassical diffraction effects are found. One consequence is that a preliminary transformation of the scattering amplitude is required before applying the Poisson representation to obtain fast convergence.

This transformation, first employed by Debye (1908) in scattering by a circular cylinder, is analogous to the multiple internal reflection treatment of the Fabry–Perot interferometer (Born & Wolf 1959). In a geometrical-optic approximation, it corresponds to the ray-tracing procedure illustrated in fig. 2.1: the scattering amplitude is decomposed into an infinite series of terms representing the effects of successive internal reflections.

In the present chapter, we discuss the results obtained by the CAM method for the first two terms of this series, that represent the effects of direct reflection from the surface and direct transmission through the sphere, employing transitional approximations (Nussenzveig 1969a, Khare 1975). One new feature found is the penetration of diffracted rays within the scatterer, taking 'shortcuts' through it before reemerging as surface waves.

## 9.1 The effective potential

The Mie scattering amplitudes were given in Sec. 5.1. Unless otherwise stated, we assume that the sphere is perfectly transparent, so that the relative refractive index $N$ is real. This is the hardest case to treat: absorption damps out contributions from all but the shortest paths through the sphere and tends to greatly improve convergence; the CAM treatment

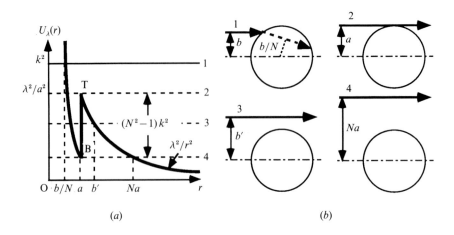

(a)                                                      (b)

Fig. 9.1. *(a)* Effective potential for transparent sphere, showing four 'energy levels'.
*(b)* Corresponding incident rays and impact parameters. In situation 1, the first
refracted ray, passing at a distance *b/N* from the center, is also shown.

remains valid for complex $N$. We assume that $N > 1$, reserving the
discussion of $N < 1$ for Chapter 12.

The effective potential in this problem can be defined through the
radial equation obeyed by the Debye electromagnetic potentials (Born &
Wolf 1959). By (1.9), at a given wave number $k$, a transparent sphere
with $N > 1$ corresponds to a square potential well with the same radius
and well depth $(N^2 - 1)k^2$, in units $\hbar = 2m = 1$. The effective potential
$U_\lambda(r)$ is the sum of this well with the centrifugal potential $\lambda^2/r^2$.

For perpendicular polarization $(j = 1)$, the $S$-function for the Mie
problem is identical to that for nonrelativistic quantum scattering by the
corresponding square well (Nussenzveig 1969a); for parallel polarization
$(j = 2)$, the effective potential must be employed with different boundary
conditions at $r = a$, yielding a different $S$-function.

The effective potential $U_\lambda(r)$ is illustrated in fig. 9.1$(a)$ for a fixed
angular momentum $\lambda$ and for four values of the 'energy' $k^2$ at different
levels (the variation of the well depth with $k^2$ is disregarded to simplify the
figure). By the localization principle, these 'energy levels' correspond to
four different impact parameters $\lambda/k$, represented in fig. 9.1$(b)$.

The effective potential has the shape of a potential pocket
surrounded by a cusp-like barrier. Situation 1 in fig. 9.1 corresponds to a
below-edge impact parameter $b < a$, for which $\beta = ka > \lambda = kb$. The
incident particle penetrates within the potential pocket and is turned back
when it hits the barrier. Fig. 9.1$(b)$ shows the first refracted ray, which

passes at a distance $b/N$ from the center, corresponding to the radial turning point.

Situation 2, $\beta = \lambda$, corresponds to an incident *edge ray* [cf. (8.1)]. The corresponding energy level sits right at the top T of the cusped barrier. In situation 3, $\beta < \lambda$, we have an above-edge ray, whose impact parameter $b'$ coincides with the radial turning point, where the centrifugal barrier is hit.

Finally, in situation 4, $\beta = \lambda/N$ (i.e., $\lambda = \alpha \equiv N\beta$), the energy level is at the bottom B of the potential pocket in fig. 9.1. In the domain between T and B, the effective potential has the shape typically associated with the formation of sharp resonances.

## 9.2   Regge poles

By (5.8), the $S$-function in the $\lambda$ plane is given by

$$S^{(j)}(\lambda,\beta) = -\frac{H_\lambda^{(2)}(\beta)}{H_\lambda^{(1)}(\beta)} \left( \frac{\{2\beta\} - Ne_j\{\alpha\}}{\{1\beta\} - Ne_j\{\alpha\}} \right) \tag{9.1}$$

where

$$\{jz\} \equiv \ln' H_\lambda^{(j)}(z) + \tfrac{1}{2} z^{-1}, \qquad \{z\} \equiv \ln' J_\lambda(z) + \tfrac{1}{2} z^{-1} \tag{9.2}$$

The Regge poles are the roots of

$$\{1\beta\} - Ne_j\{\alpha\} = 0 \tag{9.3}$$

The Regge pole distribution for a given polarization is schematically represented in fig. 9.2. For a square potential well, details of the distribution have been discussed by several authors (Bollini & Giambiagi 1962, 1963, Barut & Calogero 1962, Patashinskii, Pokrovskii & Khalatnikov 1963, Nussenzveig 1969a). For Mie scattering, the distribution for each polarization is similar to that for the potential well (Khare 1975).

The poles fall into two quite different classes. One of them, located in regions 4 and 5 (fig. 9.2), very close to the Regge poles for an impenetrable sphere (cf. fig. 8.2), is associated with surface waves that are quite similar to those found in that problem. These waves are relatively insensitive to the shape of the effective potential for $r < a$: their

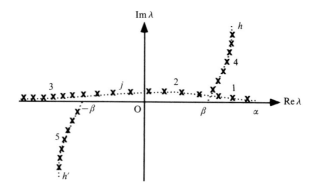

Fig. 9.2. Regge poles for $N > 1$ and a given polarization. **x** poles; 1 narrow resonances; 2 broad resonances; 4 surface waves.

behavior is determined almost entirely by the geometry of the surface.

The other class of poles, distributed above the real axis along a curve $j$ that extends to $\lambda = \alpha$, is affected by the potential in the internal region. The poles located in region 1, between $\lambda = \beta$ and $\lambda = \alpha$, are associated with the narrow resonances formed in the domain between T and B in fig. 9.1(a). They may approach very close to the real axis. These poles and their physical effects will be discussed in Chapter 14.

The poles in region 2, to the left of $\lambda = \beta$, are farther from the real axis, corresponding to broader resonances above the top of the cusped barrier in fig. 9.1(a). To the left of the imaginary axis, in region 3, the poles tend to approach the negative integers.

The separation between the poles along the curve $j$ is of order unity, so that the number of resonance-type poles in the first quadrant is $O(\beta)$. Since their imaginary parts are not large, a residue series over these poles would converge about as slowly as the partial-wave series. Thus, it would not be advantageous to apply the Poisson sum formula directly to the total scattering amplitude (5.6).

## 9.3   The Debye expansion

The convergence difficulties found, in contrast with the impenetrable sphere problem, when one tries to apply the CAM method directly to the total amplitude arise from the penetration of the waves inside the sphere. In the geometrical-optic limit, as illustrated in fig. 2.1, the interaction with the sphere is described as a sequence of interactions with its surface

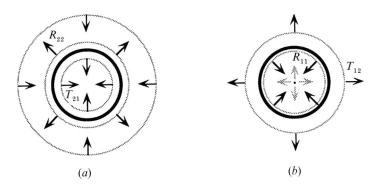

Fig. 9.3. *(a)* An incoming spherical wave is partially reflected and partially transmitted. *(b)* The transmitted spherical wave goes through the center and back to the surface, where it undergoes partial internal reflection and partial transmission.

(partial reflections and transmissions), alternating with propagation inside it. To overcome the convergence difficulties, we now imitate this procedure by looking for a description in terms of successive surface interactions.

To isolate the interaction of a spherical multipole wave with the surface, we have to treat a fictitious problem in which the radial coordinate ranges from $-\infty$ to $\infty$ and the surface separates two unbounded homogeneous media: medium 1 (the interior of the sphere) and medium 2 (the exterior region). This allows us to obtain the *spherical reflection and transmission coefficients* (Nussenzveig 1969a).

As illustrated in fig. 9.3*(a)*, an incoming spherical multipole wave with angular momentum $\lambda$ in medium 2 is partially reflected back into medium 2, defining the *external spherical reflection coefficient* $R_{22}^{(j)}(\lambda,\beta)$ (where $j$ is the polarization index), and partially transmitted into medium 1, defining the *spherical transmission coefficient* from 2 to 1, $T_{21}^{(j)}(\lambda,\beta)$. It is found, for example, employing the same notation as in (9.1), that

$$R_{22}^{(j)}(\lambda,\beta) = -\frac{\{2\beta\} - Ne_j\{2\alpha\}}{\{1\beta\} - Ne_j\{2\alpha\}} \qquad (9.4)$$

The transmitted incoming wave goes through the center of the sphere, which converts it into an outgoing wave, behaving like a perfect reflector (in view of the regularity condition at the origin). When this outgoing wave hits the interface [fig. 9.3*(b)*], it undergoes partial reflection, defining the *internal spherical reflection coefficient* $R_{11}^{(j)}(\lambda,\beta)$, and partial transmission to the outside medium, defining the *spherical* *transmission coefficient* from 1 to 2, $T_{12}^{(j)}(\lambda,\beta)$. The reflected incoming wave goes through a similar cycle, so that we get an infinite series the terms of which represent the results of successive internal reflections.

The corresponding *Debye expansion of the S-function* is of the form (Nussenzveig 1969a, Khare 1975, Fiedler-Ferrari, Nussenzveig & Wiscombe 1991)

$$S^{(j)}(\lambda,\beta) = S_0^{(j)}(\lambda,\beta) + \sum_{p=1}^{P} S_p^{(j)}(\lambda,\beta) + \Delta S_P^{(j)}(\lambda,\beta), \qquad j = 1,2 \quad (9.5)$$

where $P$ is the order of the last term that one wants to retain and

$$S_0^{(j)}(\lambda,\beta) = \frac{H_\lambda^{(2)}(\beta)}{H_\lambda^{(1)}(\beta)} R_{22}^{(j)}(\lambda,\beta) \qquad (9.6)$$

$$S_p^{(j)}(\lambda,\beta) = \frac{H_\lambda^{(2)}(\beta)}{H_\lambda^{(1)}(\beta)} T_{21}^{(j)}(\lambda,\beta) T_{12}^{(j)}(\lambda,\beta) \left[\rho^{(j)}(\lambda,\beta)\right]^{p-1} \frac{H_\lambda^{(1)}(\alpha)}{H_\lambda^{(2)}(\alpha)}$$
$$(p = 1,2,\ \ldots) \quad (9.7)$$

$$\Delta S_P^{(j)}(\lambda,\beta) = S_{P+1}^{(j)}(\lambda,\beta) \Big/ \left[1 - \rho^{(j)}(\lambda,\beta)\right] \qquad (9.8)$$

with

$$\rho^{(j)}(\lambda,\beta) = \frac{H_\lambda^{(1)}(\alpha)}{H_\lambda^{(2)}(\alpha)} R_{11}^{(j)}(\lambda,\beta) \qquad (9.9)$$

The first term (9.6) is associated with direct reflection. The term of order $p$ (9.7) corresponds to transmission after $(p - 1)$ internal reflections (direct transmission, for $p = 1$). The *remainder* of the expansion is given by (9.8). The ratio $\rho^{(j)}$ between successive terms in the geometric series contains, besides the internal reflection coefficient, the phase factor $H_\lambda^{(1)}(\alpha)/H_\lambda^{(2)}(\alpha)$, associated with propagation to the center and back. The term of order $p$ represents the effect of $(p + 1)$ surface interactions.

Substituting (9.5) into (5.6), we obtain the *Debye expansion of the total scattering amplitudes*

$$S_j(\beta,\theta) = S_{j0}(\beta,\theta) + \sum_{p=1}^{P} S_{jp}(\beta,\theta) + \Delta S_{jP}(\beta,\theta) \qquad (9.10)$$

where $S_{j0}$ is the *direct reflection term*, $S_{j1}$ is the *direct transmission term*, and $S_{jp}$ ($p \geq 2$) corresponds to transmission following ($p - 1$) internal reflections. The Poisson sum formula is now applied to each term of the expansion in order to obtain rapidly convergent results, as discussed in the following section.

## 9.4 Convergence of the Debye expansion

The numerical convergence of the Poisson-transformed Debye expansion involves two quite different aspects: how fast the asymptotic representation of each term, given by the CAM method, converges; and how fast the Debye expansion itself converges.

The Poisson representation of each term in the expansion yields asymptotic approximations by following the general procedure outlined in Sec. 7.4: path deformations in the $\lambda$ plane lead to background integrals, usually dominated by saddle point contributions, and there are also contributions from residues at complex poles, associated with the path deformations (all the integrands are meromorphic functions).

The poles arise from the denominator of (9.4), which is common to all of the spherical reflection and transmission coefficients, so that they are the roots of

$$\{1\beta\} - Ne_j\{2\alpha\} = 0 \tag{9.11}$$

which differs from (9.3) by the substitution $\{\alpha\} \to \{2\alpha\}$, corresponding to the transition from standing to ingoing waves within the sphere. We refer to them as *Regge–Debye poles*. Although they are the same for all terms of the Debye expansion, their order varies: they are of order $p + 1$ for the $p$th term ($p = 0, 1, 2, ...$).

The distribution of Regge–Debye poles for a given polarization (Streifer & Kodis 1964, Nussenzveig 1969a, Khare 1975) is schematically represented in fig. 9.4. It is symmetrical with respect to the origin, so that it suffices to consider the upper half-plane. No 'resonance-type' poles near the real axis, such as those in fig. 9.2, are found. The poles $\lambda_n^{(j)}$ in the first quadrant are all of 'surface-wave type', and they are still close to the positions $\beta + e^{i\pi/3}x_n/\gamma$ found for an impenetrable sphere [cf. (7.17)]. There is a new set of poles $\lambda'{}_n^{(j)}$ in the second quadrant, located near the points $-\alpha + e^{2i\pi/3}N^{1/3}x_n/\gamma$, with imaginary parts larger than those of the corresponding poles in the first quadrant. Thus, in contrast with the Regge poles of fig. 9.2, we expect that residue series contributions from the

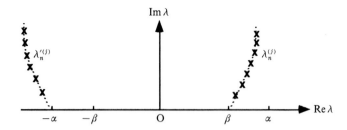

Fig. 9.4. Regge–Debye pole distribution for $N > 1$ and a given polarization $j$.

Regge–Debye poles are rapidly damped as $n$ increases. It follows that *the CAM method leads to rapidly convergent asymptotic approximations for each term of the Debye series.*

The absence of resonance effects in individual Debye terms and the presence only of surface-wave type pole contributions agree with the physical interpretation of the Debye expansion as a description based exclusively on surface interactions.

The other question that has to be discussed is how fast the Debye series itself converges. While a fuller discussion of this question must wait until we have developed further insight into the behavior of the various terms, we can anticipate some of the results.

According to (9.7), successive terms of the Debye expansion differ by an additional factor $\rho^{(j)}(\lambda,\beta)$ in their Poisson transform integrands. For real $\lambda$ and $N$, by (9.9),

$$\left| \rho^{(j)}(\lambda,\beta) \right| = \left| R_{11}^{(j)}(\lambda,\beta) \right| \tag{9.12}$$

so that the rate of convergence depends on the magnitude of the internal spherical reflection coefficient within the $\lambda$ range that yields significant contributions.

Contributions from real saddle points are associated with geometrical-optic rays, for which the Debye terms are in one-to-one correspondence with the scattered rays 0, 1, 2, ... shown in fig. 2.1. The angle of incidence $\theta_1$ in this figure is related to the saddle point $\overline{\lambda}$ by

$$\sin \theta_1 = b/a = \overline{\lambda}/\beta \tag{9.13}$$

where we have employed the localization principle. The rate of convergence of real saddle-point contributions is the same as that of the multiple internal reflection geometrical-optic series, determined by the Fresnel internal reflection coefficient at the interface.

Except for $N \gg 1$, the reflection coefficient is fairly small over nearly the whole range of angles of incidence (Born & Wolf 1959), leading to rapid damping by successive internal reflections; if absorption is present, of course, it produces even faster damping. We have already seen in connection with the rainbow that the secondary bow is much fainter than the primary one because of the extra internal reflection. For water droplets, over 98.5% of the total intensity goes into the geometrical-optic contributions to the first three terms of the Debye expansion (0, 1, 2), so that for many purposes it suffices to consider only these terms (van de Hulst 1957).

Even though only a small fraction of the total intensity goes into higher-order Debye terms, this does not prevent it from becoming highly concentrated into some very narrow angular neighborhood of special directions, as will be seen to happen in the glory, or of special size parameters (resonances). However, to survive the effects of internal reflection damping, such higher-order contributions must be nearly totally internally reflected. This happens near glancing incidence, i.e., for *near-edge rays.*

The corresponding $\lambda$ domain is the *edge strip* (8.3)–(8.4), which encompasses 'energy levels' lying near the top T of the barrier in fig. 9.1. Since surface-wave contributions are also launched by near-edge tunneling (Sec. 8.7), one should expect the convergence rate to be slow for residue-series Debye contributions. For above-edge incidence, the difference between the reflection coefficient (9.12) and unity becomes exponentially small as $\lambda$ increases, so that the Debye expansion should not be employed in the treatment of resonances (Chapter 14).

## 9.5  Direct reflection term

The first term $S_{j0}(\beta,\theta)$ of the Debye expansion (9.10), associated with direct reflection from the surface of the sphere, differs from the hard-sphere scattering amplitude only by the substitution of (7.20) by (9.6), so that the hard-sphere reflection coefficient ($-1$) is replaced by the external Mie spherical reflection coefficient $R_{22}^{(j)}(\lambda,\beta)$, given by (9.4). Thus, the structure of this term, for $N > 1$, is very similar to that discussed in Chapter 8, and it suffices to mention some of the main differences.

In the geometrical-reflection region $\theta \gg \gamma$, the amplitude is dominated by the contribution from the geometrical reflection saddle point $\bar{\lambda} = \beta \cos(\theta/2)$ [cf. (7.21)], which is far enough from $\lambda = \beta$ so that the

Debye asymptotic expansions of the cylindrical functions (Abramowitz & Stegun 1965) can be employed in (9.4), yielding

$$R_{22}^{(j)}(\lambda,\beta) \approx \frac{\left(\beta^2 - \lambda^2\right)^{\frac{1}{2}} - e_j\left(\alpha^2 - \lambda^2\right)^{\frac{1}{2}}}{\left(\beta^2 - \lambda^2\right)^{\frac{1}{2}} + e_j\left(\alpha^2 - \lambda^2\right)^{\frac{1}{2}}} \tag{9.14}$$

The dominant saddle-point contribution is the zero-order WKB result [cf. (8.15)]

$$S_{j0}^{\text{WKB}}(\beta,\theta) \approx -\frac{i\beta}{2}\left\{\frac{\sin(\theta/2) - e_j\left[N^2 - \cos^2(\theta/2)\right]^{\frac{1}{2}}}{\sin(\theta/2) + e_j\left[N^2 - \cos^2(\theta/2)\right]^{\frac{1}{2}}}\right\}$$
$$\times \exp\left[-2i\beta\sin(\theta/2)\right] \tag{9.15}$$

where the expression within curly brackets, obtained by taking (9.14) at the saddle point, is the *Fresnel reflection amplitude* at a plane interface (Born & Wolf 1959) for polarization $j$ and the angle of incidence $\theta_1 = \frac{1}{2}(\pi - \theta)$ (see fig. 7.1). Thus, in the WKB approximation, the spherical reflection coefficient is replaced by the Fresnel one. This amounts to neglecting the curvature of the spherical surface, substituting it by the tangent plane at the point of incidence, as is typical of geometrical optics.

The residue-series arising from the Regge–Debye poles $\lambda_n^{(j)}$ of fig. 9.4 is very similar to that found in (8.18), with the same physical interpretation: it represents surface-wave contributions that only tend to become appreciable as one approaches the penumbra region.

Contributions from the other set of Regge–Debye poles $\lambda'_n^{(j)}$ are always negligible, because their residues contain an exponentially small factor of the type $\exp(-c\beta)$ with $c > 0$ (Nussenzveig 1969a, Khare 1975).

Going over to the penumbra region, $\theta = O(\gamma)$, the blocking amplitude (which, as we have seen, does not depend on the internal structure of the scatterer), remains dominant within the forward diffraction peak, $\theta = O(\beta^{-1})$, corresponding to the classical Airy pattern [cf. (8.20)]:

$$S_{j0,b}(\beta,\theta) \approx \beta\left(\frac{\theta}{\sin\theta}\right)^{\frac{1}{2}} \frac{J_1(\beta\theta)}{\theta} \tag{9.16}$$

For water droplets, this gives rise to the corona. There are Fock-type

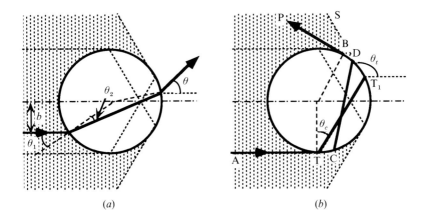

Fig. 9.5. *(a)* Directly transmitted ray; *(b)* Diffracted rays TT$_1$BP and TCDBP.

corrections, similar to those found in Sec. 8.5, but refractive-index dependent (see Chapter 13).

For larger diffraction angles within the penumbra region, $\theta \gtrsim \gamma$, transitional asymptotic approximations of Fock type have been developed (Nussenzveig 1969a, Khare 1975). A uniform approximation, very similar to that discussed in Chapter 8, is required to obtain greater accuracy and a broader domain of validity, leading to a smooth match with the geometrical-reflection region.

## 9.6 Direct transmission term

The second term $S_{j1}(\beta,\theta)$ of the Debye expansion (9.10) is associated with direct transmission through the sphere, without any internal reflection; at the geometrical-optic level, this corresponds to scattered rays of class 1 in fig. 2.1. Such rays cover an angular domain bounded by the limiting edge ray ATT$_1$S in fig. 9.5(*b*), which emerges at an angle

$$\theta_t = \pi - 2\theta_c, \qquad \theta_c \equiv \sin^{-1}(1/N) \qquad (9.17)$$

where $\theta_c$ is the critical angle for total reflection. Thus, at this level, there are two angular domains for the direct transmission term: the *1-ray (illuminated) region* $0 \le \theta < \theta_t$ and the *0-ray (shadow) region* $\theta_t < \theta \le \pi$.

Diffraction effects must occur around the geometrical shadow boundary $\theta = \theta_t$. Outside of this neighborhood, well within the illuminated region, the direct transmission term is dominated by a real saddle point in the background integral,

$$\bar{\lambda} \equiv kb = \beta \sin \theta_1 \qquad (9.18)$$

where

$$\theta_1 = \tfrac{1}{2}\theta + \theta_2 , \qquad \sin \theta_1 = N \sin \theta_2 \qquad (9.19)$$

so that $\theta_2$ is the angle of refraction corresponding to the angle of incidence $\theta_1$, and $b$ is the impact parameter of the incident ray that gets transmitted in the direction $\theta$ after two refractions, according to geometrical optics [fig. 9.5(a)].

The dominant saddle-point contribution yields the zero-order WKB approximation (Nussenzveig 1969a, Khare 1975, Fiedler-Ferrari *et al.* 1991)

$$S_{j1}^{\text{WKB}}(\beta,\theta) \approx i\beta \left[ \frac{N s_1 c_1 c_2}{2 \sin \theta (N c_2 - c_1)} \right]^{\frac{1}{2}} \frac{4 N e_j c_1 c_2}{\left(N e_j c_2 + c_1\right)^2}$$
$$\times \exp\left[2i\beta(N c_2 - c_1)\right] \qquad (9.20)$$

where

$$s_n \equiv \sin \theta_n, \quad c_n \equiv \cos \theta_n \qquad (n = 1,\, 2) \qquad (9.21)$$

The expression within the square brackets in the first line of (9.20) represents the *density of paths*, sometimes called the *beam divergence factor* (van de Hulst 1957), that accounts for the angular spreading of the initially uniform intensity distribution in an incident beam as it propagates through the scatterer [cf. the corresponding 'pre-exponential factor' in (1.28) and the Jacobian in (1.5)]. The factor that follows it is the product of the Fresnel transmission amplitudes (Born & Wolf 1959) for refraction into and out of the sphere. The phase factor in the second line of (9.20) contains the optical path difference arising from traversal of the sphere.

In the forward direction, the dominant contribution to the amplitudes is the limit of (9.20) as $\theta \to 0$:

$$S_{j1}^{\text{WKB}}(\beta,0) \approx \frac{2iN^2\beta}{(N-1)(N+1)^2} \exp\left[2i(N-1)\beta\right] \qquad (9.22)$$

The fact that this is independent of polarization is a general feature of forward scattering: for $\theta = 0$, the scattering plane is not defined and one cannot therefore distinguish between the two polarizations.

The forward scattering amplitude determines the total cross section according to the electromagnetic version (van de Hulst 1957) of the optical theorem (7.6):

$$\sigma = \left(4\pi/k^2\right)\mathrm{Re}\,S_j(\beta,0) \tag{9.23}$$

Substituting (9.16) and (9.22) into this result, we find that the dominant contributions from direct reflection and transmission to the total cross section are (neglecting Fock-type and other higher-order corrections)

$$\frac{\sigma_0 + \sigma_1}{\pi a^2} \approx 2 - \frac{8}{\beta}\,\mathrm{Im}\left\{\frac{N^2}{(N-1)(N+1)^2}\exp[2\mathrm{i}(N-1)\beta]\right\} \tag{9.24}$$

where the first term, as in (8.31), arises in equal measure from geometrical reflection and classical diffraction (blocking). The second term gives rise to a damped sinusoidal oscillation around the first one as a function of the size parameter, that represents the effect of interference with direct transmission. Further discussion of the total cross section is deferred to Chapters 13 and 14.

If $N$ is close enough to unity (which, in this context, applies even to water, $N \sim 4/3$), we see that the right-hand side of (9.24) depends almost exclusively on the parameter $\rho \equiv 2(N-1)\beta$, the optical path difference of the central ray through the sphere. Under these circumstances, paraxial rays undergo only very small deflections, and the interference between diffraction and direct transmission persists throughout the domain of small diffraction angles, giving rise to appreciable deviations from the Airy diffraction pattern. These effects, known as *anomalous diffraction*, have been investigated by van de Hulst (1946–8, 1957).

Turning now to the shadow region $\theta > \theta_t$, one finds that, well within this region, the amplitudes can be represented by rapidly converging pure residue series (Nussenzveig 1969a, Khare 1975). The residues are taken at the Regge–Debye poles, which, for the direct transmission term, are double poles. Both types of poles, $\lambda_n^{(j)}$ and $\lambda'{}_n^{(j)}$ (fig. 9.4), are included, but contributions from the poles $\lambda'{}_n^{(j)}$ are exponentially small and can be neglected.

Contributions from the poles $\lambda_n^{(j)}$ represent surface waves similar to those of Sec. 7.2, travelling from the shadow boundary into the shadow. They can be described in terms of diffracted rays generated by edge rays at glancing incidence. In fig. 9.5*(b)*, the tangentially incident ray AT gives rise to the critically refracted ray $TT_1$ that reemerges tangentially to define the shadow boundary $T_1S$ (with vanishing intensity, according to geometrical optics). At $T_1$, surface waves are launched, travelling into the shadow and shedding new rays such as BP in tangential directions.

However, before critical refraction, a diffracted ray generated at T can describe an arc along the surface, travelling as a surface wave, so that the residue series contribution in the direction of observation BP includes diffracted rays such as TCDBP in fig. 9.5*(b)* for all possible decompositions of the total arc (arc TC + arc DB) into two parts. The new effect of 'taking a shortcut' across the sphere is described in the geometrical theory of diffraction by a factor $D_{21}^{(j)}D_{12}^{(j)}$, where $D_{21}^{(j)}$ and $D_{12}^{(j)}$ represent the *transmission coefficients of surface waves* into the sphere and out of the sphere, respectively.

Within the penumbra region $|\theta - \theta_l| = O(\gamma)$, one obtains transitional approximations to the amplitudes (Nussenzveig 1969a, Khare 1975), expressed in terms of Fock-type functions. Improved results require uniform approximations similar to that developed in Chapter 8. Thus, the 1-ray/0-ray transition is a typical *Fock-type transition*, associated with the disappearance of a single real ray: in the $\lambda$ plane, one type of critical point, a real saddle point, is replaced by a set of Regge-Debye poles. As discussed in Sec. 7.3, a Fock-type function interpolates between saddle-point asymptotic behavior in an illuminated region and exponentially damped residue-series behavior in a shadow region, within a transition region with typical angular width $\gamma$.

# Theory of the rainbow

*a certain affection for this problem*
*pervades even the driest computations.*
(van de Hulst 1957)

Our review of the development of rainbow theory in Chapter 3 concluded with the remark by van de Hulst (1957) that no quantitative theory was available, apart from numerical Mie summations. The CAM method has led to an accurate quantitative theory, that will now be discussed.

The primary bow of the meteorological rainbow appears in the third term of the Debye expansion. For definiteness, we deal mainly with this term, but the treatment applies just as well to higher-order rainbows.

## 10.1  The third Debye term

The third term $S_{j2}(\beta,\theta)$ of the Debye expansion (9.10) is associated with transmission through the sphere with one internal reflection, i.e., at the geometrical-optic level, with scattered rays of class 2 in fig. 2.1. In contrast with the first two terms, it cannot be treated uniformly for all $N > 1$: already in geometrical optics, the subdivision into angular regions and the number of rays found in each region are different for different ranges of $N$ (Nussenzveig 1969b).

We consider only the range

$$1 < N < \sqrt{2} \tag{10.1}$$

because we have in mind the application to water droplets.

The deflection function in this range looks like the upper curve in fig. 3.2(*a*). As the impact parameter $b$ increases from zero, the scattering angle $\theta$ decreases from $\pi$ to a minimum value $\theta_R$, and then increases again, to a maximum value $\theta_L = 4 \cos^{-1}(1/N)$, attained for the edge ray. Thus, the domain $\theta_R < \theta < \theta_L$ is covered twice, leading to a subdivision at

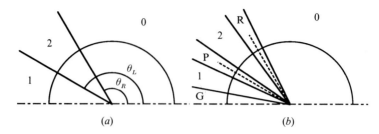

Fig. 10.1. *(a)* Geometric-optical subdivision: 0-ray (shadow), 2-ray and 1-ray
regions. *(b)* Transition regions: R (rainbow), P (penumbra) and G (glory).

the geometrical-optic level into the three angular regions shown in fig.
10.1*(a)*: the *0-ray (geometrical shadow) region* $0 \le \theta < \theta_R$; the *2-ray
region* $\theta_R < \theta < \theta_L$; and the *1-ray region* $\theta_L < \theta \le \pi$.

Diffraction effects should occur around the boundaries between
these regions, including, as will be seen later, a neighborhood of the
backward direction, so that we find three additional angular *transition
regions*, illustrated in fig. 10.1*(b)*: a *glory region* G near $\theta = \pi$, a *penumbra
region* P around the 1-ray/2-ray boundary, and a *rainbow region* R around
the 2-ray/0-ray boundary.

In each region, the Poisson-transformed amplitude is reduced to a
background integral, usually dominated by saddle-point contributions, plus
a sum of residues at Regge–Debye poles.

The saddle point positions are obtained by solving a fourth-degree
equation, with only one or two roots that are not spurious, depending on
the value of $\theta$ (Nussenzveig 1969b). The trajectories described by these
saddle points in the $\lambda$ plane as $\theta$ varies are schematically represented in
fig. 10.2.

For $\theta = \pi$, there is only one saddle point, represented by the black
circle at $\lambda = 0$, corresponding to the axial ray that penetrates into the
sphere and is reflected back. As $\theta$ decreases, this saddle point moves to
the right along the real axis. At $\theta = \theta_L$, a new saddle point, represented by
a white circle in fig. 10.2, makes its appearance at $\lambda = \beta$, associated with
the edge ray. When $\theta$ decreases below $\theta_L$, this saddle point moves along
the real axis to the left, while the former one keeps moving to the right,
until they finally meet for $\theta = \theta_R$, the rainbow angle. Thus, in the $\lambda$ plane,
*a rainbow corresponds to a collision between two saddle points.*

For $\theta < \theta_R$, the two saddle points become complex. They leave the
real axis at right angles and describe complex-conjugate trajectories, as
illustrated in fig. 10.2.

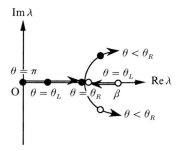

Fig.10.2. Saddle point trajectories in the $\lambda$ plane. ● ○ saddle points.

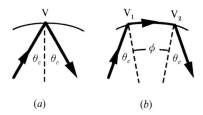

(a)                                    (b)

Fig. 10.3. The two possible types of surface interactions of diffracted rays ('vertices').

In the 1-ray region [outside of the transition regions of fig. 10.1(b)], the dominant contribution to the amplitude, arising from the single real saddle point, corresponds to the WKB approximation and is analogous to (9.20), including one additional factor representing the internal Fresnel reflection coefficient.

The residue contribution from the Regge–Debye poles $\lambda_n^{(j)}$ of fig. 9.4 (which are triple poles for this term) corresponds, in the language of the geometrical theory of diffraction, to diffracted rays excited at edge incidence, which take *two* shortcuts of the type illustrated in fig. 9.5(b) across the sphere and interact internally once with the surface in between them.

This surface interaction can be just an internal total reflection, with the Fresnel reflection coefficient $R_{11}^{(j)} = 1$ associated with internal incidence at the critical angle $\theta_c$, as illustrated in fig. 10.3(a), but it can also be of the type shown in fig. 10.3(b): the diffracted ray is critically refracted to the outside, travels an angle $\phi$ as an evanescent wave along the surface, and undergoes a new critical refraction to the inside. This type of interaction gives rise to a factor $D_{21}^{(j)}D_{12}^{(j)}$, like that found in Sec. 9.6.

The total surface-wave contribution in a given direction may be

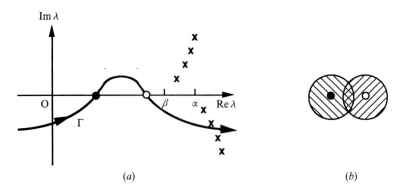

*(a)*                                                    *(b)*

Fig. 10.4. *(a)* Path of integration Γ in 2-ray region. ● ○ saddle points. ✗ Regge–Debye
poles. *(b)* Saddle points with overlapping ranges.

represented by a series of diagrams in which these two types of
interaction play the role of 'vertices' (Nussenzveig 1969b). They must be
combined in all possible ways that lead to taking two shortcuts. An
example is the surface-wave contribution to the glory represented in fig.
4.3*(b)*, where BC corresponds to a vertex of the type shown in fig.
10.3*(b)*.

The structure of the amplitudes in the 2-ray region, outside of the
transition regions, is similar to that found in the 1-ray region, except for
the presence of an additional real saddle point in the background integral.
As shown in fig. 10.4*(a)*, the path of integration Γ in this integral must be
taken over both saddle points, yielding the sum of the WKB terms
associated with the corresponding geometrical-optic rays. These are the
two rays that interfere on the bright side of the rainbow, and *the WKB
approximation corresponds to Young's interference theory of the rainbow,*
corrected for the phase change across focal lines (cf. Sec. 3.2).

The 1-ray/2-ray transition region, $|\theta - \theta_L| = O(\gamma)$, denoted as P in
fig. 10.1*(b)*, is a typical Fock-type penumbra region associated with the
disappearance of a single real ray, with characteristic angular width of
order $\gamma$.

The 0-ray region, corresponding to the rainbow shadow outside of
the rainbow transition region, differs from the shadow region found in Sec.
9.6 because the amplitudes cannot be reduced to pure residue series at
the Regge–Debye poles. The background integral remains present, and
one can verify (Nussenzveig 1969b) that it must be taken only over the
lower complex saddle point shown in fig. 10.2.

The corresponding saddle-point contribution is the analytic
continuation of a real saddle-point WKB contribution found in the 2-ray

region: rather than oscillatory, however, it is damped faster than exponentially as one penetrates into the shadow, with a damping factor proportional to

$$\exp\left[-c\left(|\theta - \theta_R|/\gamma^2\right)^{3/2}\right]$$

where $c$ is a constant. This term can be interpreted as a *complex ray*, a generalization of real geometric-optic rays to complex propagation constants, typically found on the shadow side of caustics (Keller 1958).

Since the angular penetration depth associated with the above damping factor is of order $\gamma^2$, while it is of order $\gamma$ for the surface-wave contributions [cf. (7.9)], the residue contributions from Regge–Debye poles eventually become dominant well within the shadow, whereas the complex saddle-point contribution becomes dominant as one approaches the penumbra region.

## 10.2   The Chester-Friedmann-Ursell method

As shown in fig. 10.2, the rainbow angle is characterized, in the $\lambda$ plane, by the confluence of the two saddle points associated with geometrical-optic rays in the 2-ray region. What happens in the saddle-point method when two saddle points approach one another?

Consider the complex integral

$$F(\kappa,\varepsilon) = \int g(w)\exp[\kappa f(w,\varepsilon)]\mathrm{d}w \qquad (10.2)$$

over some path in the $w$ plane, where $\kappa$, the asymptotic expansion parameter (like $\beta$ in our case), is large and positive and $\varepsilon$ (like $\theta$ in our case) is an independent parameter. If, for some value of $\varepsilon$, as happens in the 1-ray region, the integral is dominated by a single saddle point $\overline{w} = \overline{w}(\varepsilon)$, around which $f$ and $g$ are regular, one can approximate, in this neighborhood,

$$f(w,\varepsilon) \approx f(\overline{w},\varepsilon) + \tfrac{1}{2} f_w''(\overline{w},\varepsilon)(w - \overline{w})^2 \qquad (10.3)$$

where $f_w''$ denotes the second derivative with respect to $w$.

By choosing the steepest-descent path through $\overline{w}$, the relevant portion of the integral becomes of Gaussian type. This means that the *range* of the saddle point, i.e., the radius of the complex neighborhood around it that yields the dominant contribution to the integral (de Bruijn 1958), has an order of magnitude given by

$$\Delta w \sim \kappa^{-\frac{1}{2}} \left| f_w''(\overline{w},\varepsilon) \right|^{-\frac{1}{2}} \tag{10.4}$$

The asymptotic expansion of the integral can be obtained by a change of variable that transforms the exponent into an exact Gaussian, followed by a power series expansion of $g$ around the saddle point and integration term by term (Olver 1974).

If $\varepsilon$ is such that two saddle points, $\overline{w}'$ and $\overline{w}''$, contribute along the path (as happens in the 2-ray region), their contributions can be added independently, *so long as their ranges do not overlap*. When the ranges overlap, as in fig. 10.4(*b*), the usual saddle-point method breaks down. Indeed, the radius of convergence of the expansion of $g$ is of the order of the distance to the nearest saddle point, where another zero of $f_w''(w,\varepsilon)$ is located. The confluence of the two saddle points takes place at a value of $\varepsilon$, say $\varepsilon = 0$, such that $f_w'' = 0$, so that the quadratic term in (10.3) is replaced by a cubic term.

For each *fixed* $\varepsilon \neq 0$, we can still add up the two independent saddle point contributions, provided that their ranges are small enough not to overlap, i.e., by (10.4), provided that $\kappa$ is large enough. However, as $\varepsilon$ varies, approaching $\varepsilon = 0$, we must take $\kappa \to \infty$, so that the saddle point result is not valid *uniformly* in $\varepsilon$. Similarly, we can obtain an asymptotic expansion in a neighborhood of $\varepsilon = 0$ by expanding $f$ up to the cubic term around the point of coalescence $\overline{w}(\varepsilon = 0)$, but this is a *transitional* approximation, again not uniformly valid in $\varepsilon$.

An extension of the method of steepest descents that yields a *uniform asymptotic expansion* was formulated by Chester, Friedman & Ursell (1957) and further extended by Ursell (1965). We refer to it as the CFU method.

The basic idea is to transform the exponent of (10.2) into an *exact* cubic through a change of variables:

$$f(w,\varepsilon) = \tfrac{1}{3}u^3 - \zeta(\varepsilon)u + A(\varepsilon) \tag{10.5}$$

with the two saddle points transformed into $\pm\zeta^{\frac{1}{2}}(\varepsilon)$. The crucial point in this as well as other uniform approximation techniques (Berry & Mount 1972) is to introduce a *mapping that preserves the saddle point structure*.

With the correct choice of branch determined by the saddle point correspondence, it can be shown that the transformation $w \leftrightarrow u$ is regular and one-to-one near $u = 0$. By substitution into (10.2) and suitable expansion of the integrand, one obtains an asymptotic expansion of the form

$$F(\kappa,\varepsilon) = \exp[\kappa A(\varepsilon)]$$

$$\times \left\{ \kappa^{-\frac{1}{3}} \left[ \sum_{s=0}^{M-1} a_s(\varepsilon)\kappa^{-s} + O(\kappa^{-M}) \right] \mathrm{Ai}[\kappa^{\frac{2}{3}}\zeta(\varepsilon)] \right.$$

$$\left. + \kappa^{-\frac{2}{3}} \left[ \sum_{s=0}^{M-1} b_s(\varepsilon)\kappa^{-s} + O(\kappa^{-M}) \right] \mathrm{Ai}'[\kappa^{\frac{2}{3}}\zeta(\varepsilon)] \right\} \quad (10.6)$$

where Ai is the Airy function (3.2) and the CFU coefficients $a_s$ and $b_s$ are regular functions of $\varepsilon$ that can be obtained from the expansion of the integrand after the transformation.

For $|z| \gg 1$ and $|\arg z| < \pi$, we have (Abramowitz & Stegun 1965)

$$\mathrm{Ai}'(z) \approx -z^{\frac{1}{2}}\mathrm{Ai}(z) \quad (10.7)$$

so that the Ai$'$ corrections in (10.6) can become of the same order as the Ai terms. If suitable regularity conditions are satisfied (Ursell 1965), the domain of applicability of the uniform asymptotic expansion (10.6) can be enlarged by matching it with the steepest-descent expansion, with which it has a common region of validity. Thus, in contrast with transitional asymptotic approximations, the uniform expansion yields a smooth match with the saddle-point results.

## 10.3  Uniform CAM rainbow approximation

In the application of the CFU method to the CAM theory of the rainbow (Nussenzveig 1969b, Khare & Nussenzveig 1974, Khare 1975), the parameters found in (10.2) are identified as

$$\kappa \equiv 2\beta, \qquad \varepsilon \equiv \theta - \theta_R \quad (10.8)$$

where $\theta_R$ is the rainbow scattering angle. The saddle points, in the 2-ray region, are [cf. (9.18)]

$$\bar{\lambda}' = \beta \sin \theta_1', \qquad \bar{\lambda}'' = \beta \sin \theta_1'' \quad (10.9)$$

where $\theta_1'$, $\theta_1''$ are the angles of incidence of the two geometrical-optic rays that merge at the rainbow angle. Denoting by $\theta_2'$, $\theta_2''$ the corresponding angles of refraction, it is found that, in (10.6),

$$\left\{\begin{array}{c} A(\varepsilon) \\ \frac{2}{3}[\zeta(\varepsilon)]^{3\!/\!2} \end{array}\right\} = \tfrac{1}{2}\mathrm{i}\big[(2N\cos\theta_2' - \cos\theta_1') \pm (2N\cos\theta_2'' - \cos\theta_1'')\big] \quad (10.10)$$

are related, respectively, with half the sum and half the difference of the optical paths through the sphere.

Finally, the dominant contribution to the third Debye term in the rainbow region is found to be of the form

$$S_{j2}(\beta,\theta) \approx \beta^{7/6}\exp[2\beta A(\varepsilon)]$$
$$\times\Big\{\big[c_{j0}(\varepsilon) + \beta^{-1}c_{j1}(\varepsilon) + \cdots\big]\mathrm{Ai}\big[(2\beta)^{2\!/\!3}\zeta(\varepsilon)\big]$$
$$+\big[d_{j0}(\varepsilon) + \beta^{-1}d_{j1}(\varepsilon) + \cdots\big]\beta^{-1\!/\!3}\mathrm{Ai}'\big[(2\beta)^{2\!/\!3}\zeta(\varepsilon)\big]\Big\} \quad (10.11)$$

where the CFU coefficients $c_{jn}(\varepsilon)$ and $d_{jn}(\varepsilon)$ can be expressed in terms of the Fresnel reflection and transmission coefficients and their derivatives of various orders with respect to the angle of incidence, evaluated at the exactly known saddle points $\theta_1'$, $\theta_1''$ (the order of the derivatives increases with $n$).

For $|\varepsilon| \ll \gamma$, expansions into power series in $\varepsilon$ can be employed: in particular,

$$\zeta(\varepsilon) = \mathrm{e}^{-\mathrm{i}\pi}(3s)^{-1\!/\!3}c\varepsilon + O(\varepsilon^2) \quad (10.12)$$

where

$$c = \big[\tfrac{1}{3}(N^2 - 1)\big]^{1\!/\!2}, \qquad s = \big[\tfrac{1}{3}(4 - N^2)\big]^{1\!/\!2} \quad (10.13)$$

are, respectively, the cosine and the sine of the rainbow angle of incidence. The $\varepsilon$-expansions of the first few CFU coefficients for $N = 1.33$ are tabulated in Khare & Nussenzveig 1974. One finds that, at $\varepsilon = 0$,

$$|c_{20}| \approx |c_{10}|/5 \quad (N = 1.33) \quad (10.14)$$

which tends to suppress polarization 2. The reason for this suppression is that the rainbow angle of incidence happens to be close to Brewster's angle, and $c_{j0}$ is proportional to the Fresnel reflection coefficient.

The classical Airy theory of the rainbow (Sec. 3.2) is obtained from (10.11) by neglecting all terms in Ai′, retaining only $c_{j0}$ in the coefficient of Ai and taking it at $\varepsilon = 0$, and neglecting all terms of order $\varepsilon^2$ and higher in the expansions of $A(\varepsilon)$ and $\zeta(\varepsilon)$.

## 10.4 CAM rainbow theory predictions

The uniform CAM rainbow approximation (10.11) confirms some results of the Airy theory while leading to different predictions for some features of the rainbow.

(i) *Rainbow enhancement.* Typical geometric-optic contributions to the scattering amplitudes [cf. (9.15), (9.20)] are $O(\beta)$, whereas the rainbow terms (10.11) are $O(\beta^{7/6})$ in the rainbow region. Thus, in agreement with the Airy theory, *the maximum rainbow intensity enhancement is of order $\beta^{1/3}$*.

(ii) *Rainbow width.* The main rainbow peak (cf. fig. 3.4) is the domain where the argument of the Airy functions in (10.11) remains of order unity. Thus, according to (10.12), *the angular width of the rainbow region is of order $\gamma^2$*. This is much narrower than the typical penumbra angular width, of order $\gamma$. Combining (i) and (ii), we see that only a fraction of the total scattered intensity of order $\gamma$ goes into a rainbow.

(iii) *Rainbow polarization.* According to (10.14), $i_2 \sim 0.04 i_1$ at the rainbow angle, so that the rainbow, already in the Airy theory, has a degree of polarization [cf. (5.5)] of about 96%, with almost complete dominance of perpendicular polarization, i.e., of light with the electric field vector directed tangentially to the rainbow arc, in agreement with observations (Volz 1961, Können 1985). However, the numerical value of the degree of polarization will vary with the angle of observation and the size parameter, and in this respect (10.11) predicts results that differ from Airy's theory.

(iv) *Angular distribution and supernumerary peaks.* Since the CFU expansion is rapidly convergent (Khare & Nussenzveig 1974) for $\beta \gg 1$, the main corrections to the Airy theory arise from the Ai′ term in (10.11). In view of (10.7), these terms become of the same order in $\beta$ as the Ai terms outside of the main rainbow peak. Thus, *the validity of the Airy theory is limited, at best, to large size parameters and small deviations from the rainbow angle, within the main rainbow peak*, in agreement with van de Hulst's criticism (Sec. 3.3).

For polarization 1, the deviations from the Airy theory should be relatively small within the main peak, becoming appreciable for the supernumeraries. For polarization 2, however, large deviations are expected. Indeed, the smallness of $|c_{20}|$ in (10.11), due to the proximity to Brewster's angle, leads to *dominance of the* Ai′ *term for polarization 2* $(|d_0/c_{20}| \gg 1)$.

In particular, while the Airy theory leads to supernumerary peaks at the same angular positions for both polarizations, CAM theory predicts that *the peaks for polarization 2 will be located near the minima for polarization 1, and vice-versa*. This can readily be understood in terms of the two-ray interference that gives rise to the supernumeraries: the reflection amplitudes for polarization 2 change sign after going through zero at the Brewster angle, which happens to one of the two interfering contributions, so that constructive interference for polarization 1 becomes destructive for polarization 2. This interchange between maxima and minima for the two polarizations was observed by Bricard (1940) in a fog bow.

(v) *Uniformity.* In contrast with the Airy approximation, the CFU result (10.11) is a *uniform asymptotic expansion*. Thus, for $|\varepsilon| \gg \gamma^2$ on the bright side of the rainbow, where the separation between the two real saddle points is much larger than their range, the result goes over smoothly into the sum of their contributions, yielding the WKB approximation in the 2-ray region.

On the dark side of the rainbow, for $|\varepsilon| \gg \gamma^2$, one may employ [cf. (10.12)] the asymptotic expansion of Ai$(z)$ for large positive $z$ (Abramowitz & Stegun 1965),

$$\mathrm{Ai}(z) \approx \tfrac{1}{2}\pi^{-1/2}z^{-1/4}\exp\left(-\tfrac{2}{3}z^{3/2}\right) \qquad (z \gg 1) \qquad (10.15)$$

and one finds that the result merges smoothly with the damped complex saddle-point contribution in the 0-ray region mentioned at the end of Sec. 10.1. Thus, *diffraction into the shadow side of a rainbow takes place by tunneling*, i.e., by a complex-ray contribution.

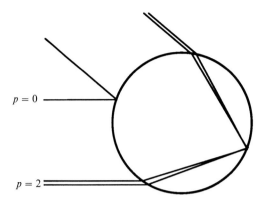

Fig. 10.5. Dominant contributions in primary rainbow region.

(vi) *Higher-order rainbows.* The uniform CAM rainbow approximation, written above for the primary bow, can be immediately extended to higher-order bows. For a Debye term of order $p$, the rainbow angle of incidence $\theta_{1R,p}$ is determined by

$$\cos\theta_{1R,p} = \left[\left(N^2 - 1\right)/\left(p^2 - 1\right)\right]^{1/2} \tag{10.16}$$

which leads to (10.13) for $p = 2$. We see that $\theta_{1R,p}$ increases monotonically with $p$, so that high-order rainbows arise from incidence in the edge strip. The angular width also increases, growing linearly with $p$ for large $p$ [cf. (11.27) below], so that the rainbow peak becomes flatter.

The weakening of higher-order rainbows by internal reflection, which is apparent for the secondary bow, becomes less pronounced as $p$ increases, because of the increased reflectivity as the edge is approached. Rainbows beyond the secondary one do not seem to be observed in the sky, because they are masked by background glare, but they are observed in the laboratory (Walker 1976, 1977, Lock 1987). However, as will be seen in the next chapter, the tenth-order rainbow is observed indirectly, through its effects on the glory.

The Fresnel reflectivities not only increase with $\theta_{1R,p}$: they also become steeper as total reflection is approached, leading to larger derivatives with respect to the angle of incidence. This enhances the contributions from higher-order CFU coefficients in (10.11), so that *the*

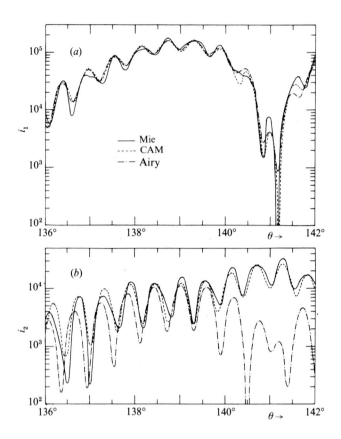

Fig. 10.6. Mie, CAM and Airy polarized intensities for $N = 1.33$, $\beta = 500$: *(a)* $i_1$; *(b)* $i_2$
(from Khare & Nussenzveig 1974).

*deviations from the Airy theory, for both polarizations, increase with p.*

## 10.5   Numerical tests

Numerical comparisons among the Mie results, the uniform CAM approximation and the Airy approximation have been performed for $N = 1.33$ and $50 \leq \beta \leq 1500$ (Khare & Nussenzveig 1974). In the primary rainbow region, the dominant Debye contributions arise from the $p = 0$ and $p = 2$ terms, as indicated in fig. 10.5.

Interference between the direct reflection ($p = 0$) contribution and

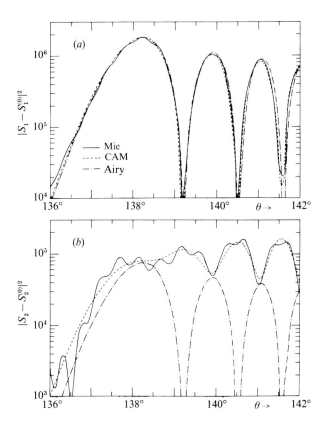

Fig. 10.7. Mie, CAM and Airy polarized intensities for $N = 1.33$, $\beta = 1500$: *(a)* $i_1$; *(b)* $i_2$. The direct reflection term has been subtracted out (from Khare & Nussenzveig 1974).

the $p = 2$ rainbow term gives rise to a rapidly oscillatory 'fine structure' superimposed on the rainbow pattern. This is apparent in fig. 10.6, which shows the polarized intensities for $\beta = 500$. Only the main rainbow peak and part of the first supernumerary appear in this figure. For polarization 1, as expected, the deviations from the Airy theory tend to become appreciable only outside of the main peak. For polarization 2, however, the out-of-phase character of the Airy approximation is quite noticeable, leading to large departures from the Mie solution.

A similar comparison for $\beta = 1500$ is shown in fig. 10.7. Here, to remove the rapid interference oscillations, the direct reflection term (9.15) has been subtracted out, so that the Mie result stands for $|S_j - S_{j,0}|^2$. Several

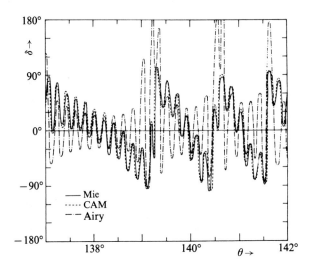

Fig. 10.8. Mie, CAM and Airy phase difference $\delta$ for $N = 1.33$ and $\beta = 1500$ (from
Khare & Nussenzveig 1974).

supernumeraries are included, and we see that all CAM predictions of
Sec. 10.4 are confirmed. For polarization 1, the deviations from the Airy
theory remain relatively small within the main peak but increase as one
goes to the supernumeraries. For polarization 2, however, the Airy
approximation breaks down, except in a small portion of the main peak:
its out-of-phase character in the supernumeraries is apparent. The small
oscillations of the Mie result around the CAM curve can be attributed to
higher-order Debye terms still present.

The phase difference $\delta$ [cf. (5.3)] for $\beta = 1500$ is plotted in fig. 10.8.
Its rapid oscillations again arise from interference with direct reflection;
large phase variations occur close to the intensity minima. Here the Airy
approximation fails even close to $\theta_R$, while the CAM result agrees with
the Mie solution remarkably well throughout.

Similar good agreement between the Mie and CAM results is
found for higher-order rainbows (cf. Fig. 11.8 below). Thus, the uniform
CAM approximation leads to an accurate quantitative theory of the
rainbow.

For natural incident light, the scattered intensity is $\frac{1}{2}(i_1 + i_2)$, so that
the dominance of polarization 1 decreases the disagreement with the Airy
theory, which can still be employed as a useful practical guide to some
features of low-order natural rainbows (Wang & van de Hulst 1991).

## 10.6   Rainbow as a diffraction catastrophe

As was mentioned in Sec. 3.3, raindrops with diameter exceeding a few tenths of a millimeter are no longer spherical: air-drag forces distort them in free fall to an oblate spheroidal shape (Volz 1961). However, such distortion of the droplet shape, as well as distortion of the incoming wavefronts from ideal planar form, does not prevent the formation of rainbows, albeit with slightly changed features. This provides experimental evidence of the *structural stability* of rainbows, i.e., their robustness under perturbations in shape or boundary conditions.

The classification of structurally stable caustics belongs to the province of *catastrophe theory* (Thom 1972, Arnold 1975, Poston & Stewart 1978). The application to caustics is reviewed by Berry & Upstill (1980).

Consider a general Huygens–Kirchhoff type integral representation of a wave field,

$$\psi(k,r) = \int A(s,r)\exp[ik\phi(s,r)]\mathrm{d}^{n}s \qquad (10.17)$$

The rays through $r$ correspond to the stationary-phase points

$$\partial\phi/\partial s_{i} = 0, \quad \forall i \qquad (10.18)$$

which defines a *gradient map* from $r$ to the space of the *state variables s*. Caustics are the *singularities* of this map, where two or more stationary-phase points coalesce. Under a perturbation of the *generating function* $\phi$, a caustic is structurally stable if this results in a smooth reversible change (diffeomorphism) in the space of *control parameters* contained in $r$.

A catastrophe is an equivalence class of structurally stable singularities. A basic result of catastrophe theory is a classification of catastrophes in terms of a set of standard polynomials (*normal forms*) for the generating function $\phi$.

In the wave representation (10.17), this leads to a corresponding set of canonical integrals called *diffraction catastrophes* (Berry & Upstill 1980). The rainbow is associated with the simplest *elementary catastrophe*, known as the *fold*, connected with the confluence of two stationary phase points. The name is related with the folding back of scattered rays that takes place at the rainbow angle. The corresponding diffraction catastrophe is the Airy function (3.2), with a cubic generating function, where $z$ plays the role of control parameter.

For each diffraction catastrophe, there are *scaling laws* and a set of characteristic power-law exponents governing its behavior in the geometrical-optic limit $k \to \infty$. The divergence of the amplitude at the caustic is governed by a power of $k$ known as the *singularity index*; for the rainbow, this is the amplitude enhancement exponent 1/6 found in Sec. 10.4. The *fringe index* (Berry & Upstill 1980) governs the scale of the diffraction fringes, which goes to zero as an inverse power of $k$. For the rainbow, as we have seen, this angular width exponent is 2/3.

Experiments with oblate spheroidal acoustically levitated water drops (Marston & Trinh 1984, Simpson & Marston 1991) show the appearance of a 'generalized rainbow' caustic that corresponds to a *hyperbolic umbilic diffraction catastrophe*. This arises from confluences involving four rays, two of which are the ordinary rainbow rays going through the circular equatorial section, while the other two are skew rays above and below the equatorial plane (Nye 1984, Marston 1985).

# Theory of the glory

*A satisfactory explanation
has not so far been available.*
(van de Hulst 1947)

CAM theory provided for the first time a detailed physical explanation of the meteorological glory. It confirmed van de Hulst's conjecture (Sec. 4.3) about the importance of surface waves of the type illustrated in fig. 4.3, but it showed that higher-order Debye contributions, both from surface waves and from the *shadow* of higher-order rainbows, are very important. What is very remarkable is that all these contributions arise from complex critical points, so that the glory is a *macroscopic tunneling effect*. Several other physical effects must be taken into account, rendering the glory one of the most complicated of all known scattering phenomena.

The meteorological glory is observed in backscattering. CAM theory predicts the existence as well of a *forward optical glory*, that will be discussed in Chapter 13.

## 11.1   Observational and numerical glory features

Observations of natural glories, together with laboratory observations, have shown some conspicuous differences between glories (sometimes called anti-coronae) and coronae (van de Hulst 1957).

(i) *Variability*. The appearance of the glory rings varies considerably from one observation to another, and sometimes even in the course of a single observation. Indeed, as will be illustrated below, the near-backward intensities are rapidly varying functions of $\beta$, $\theta$ and $N$.

(ii) *Angular distribution*. The angular distribution falls off away from the central region much more slowly than in the coronae, so that the outer rings are more pronounced: the observation of as many as five sets of rings has been reported.

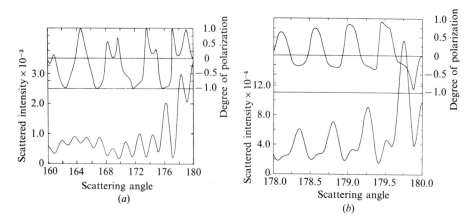

Fig. 11.1. Scattered intensity and degree of polarization in glory region, for $N = 1.342$:
(a) with $\beta = 98.2$; (b) with $\beta = 785.4$ (after Dave 1969).

(iii) *Polarization of natural glories.* For natural glories, the first dark ring tends to be rather hazy; there are also indications of strong polarization, with polarization 2 (parallel) dominant in the outer rings. According to van de Hulst (1957), the average value of the size parameter in clouds where the observations of natural glories were made is $\beta = 160$.

More precise information on glory features comes from numerical summation of the Mie series. Dave (1969) has plotted the angular distribution of the intensity for natural incident light, given by

$$i(\beta, \theta) = \tfrac{1}{2}\left[i_1(\beta, \theta) + i_2(\beta, \theta)\right] \tag{11.1}$$

and the degree of linear polarization, defined by (5.5) (which is positive or negative depending on whether polarization 1 or 2 dominates), for $N = 1.342$ and several values of $\beta$. Some of his results in the glory region are shown in fig. 11.1, where the lower curves represent the intensity and the upper ones represent the degree of linear polarization. They illustrate the following features.

(iv) *Angular width.* The angular width of the glory region is of order $\beta^{-1}$:

$$u \equiv \beta(\pi - \theta) = O(1) \tag{11.2}$$

(v) *Features around* $\beta = 10^2$. For $\beta \sim 10^2$ [fig. 11.1*(a)*], which is in the range of size parameters connected with observations of natural glories, the first dark ring is only a small depression, consistent with its observed haziness, and the outer rings tend to be parallel-polarized, again in agreement with observations.

(vi) *Features around* $\beta = 10^3$. For $\beta \sim 10^3$ [fig. 11.1*(b)*], the angular distribution is very different (showing, in this instance, a first bright ring with higher intensity than the center of the pattern), and the outer rings tend to be perpendicular-polarized.

Numerical and experimental studies of the backscattered intensity as a function of the size parameter $\beta$, for values of $\beta$ ranging from about 50 to several thousand, have been made by Bryant & Cox (1966), Fahlen & Bryant (1968), Saunders (1970), Shipley & Weinman (1978), and Ashkin & Dziedzic (1977, 1981). The following features emerge from these studies.

(vii) *Intensity enhancement.* The average backscattered intensity is, typically, at least one order of magnitude larger than that predicted by geometrical optics (cf. fig. 5.2).

(viii) *Ripple fluctuations.* Very rapid intensity fluctuations, like those illustrated in fig. 5.2, are seen for the whole range of size parameters, both numerically and in the experiments. Similar fluctuations appear in the differential cross section at other angles, as well as in the total cross section, but their relative amplitude is strongly enhanced in backscattering.

(ix) *Quasiperiodicity.* The backscattered intensity pattern is quasiperiodic in $\beta$, with a quasiperiod $\Delta\beta$ ranging from about 0.81 to about 0.83 for the range of refractive indices associated with water in the visible.

(x) *Background.* The rapid intensity fluctuations (spikes) within a quasiperiod are superimposed on a relatively slowly-varying background, that goes through about two to three humps per quasiperiod (cf. fig. 5.2).

(xi) *Effects of absorption or size averaging.* When one adds a small amount of absorption (imaginary component of the complex refractive index), the sharp spikes tend to disappear, but the background humps are not much affected. Averaging the results over a range of size parameters

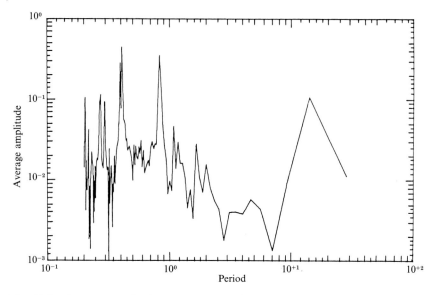

Fig. 11.2. Average amplitude spectrum as a function of period for backscattered
intensity, $N = 1.333$, $500 \leq \beta \leq 528$ (after Shipley & Weinman 1978).

produces a similar effect. Since water droplets in the atmosphere do
absorb and droplet populations have size dispersion, we see that the sharp
spikes may be disregarded in a theory of the meteorological glory: what
the theory must account for is the *background* component.

(xii) *Other periodicities.* By Fourier expanding a suitably normalized
backscattered intensity, Shipley & Weinman (1978) computed the
spectrum of the Fourier amplitude as a function of the size parameter
period. The result for $N = 1.333$ and $\beta$ near 500 is shown in fig. 11.2.
Besides the basic quasiperiod $\Delta\beta \approx 0.83$, strong periodic components with
the periods $\Delta_1\beta \approx 0.41$, $\Delta_2\beta \approx 1.1$ and $\Delta_3\beta \approx 14$ can be identified.

(xiii) *Quasichaotic features.* Besides the peaks at discrete periods, fig.
11.2 suggests the presence of a quasi-continuous background spectrum,
which would signal the presence of a chaotic component (Bergé, Pomeau
& Vidal 1987). This agrees with the 'sensitive dependence on initial
conditions' that appears to be characteristic of the ripple fluctuations
(Sec. 5.2) as well as with the observed glory variability [feature (i)].

(xiv) *Average gain factor.* The *gain factors* relative to an isotropic
scatterer are the ratios of the actual polarized intensities to their limiting

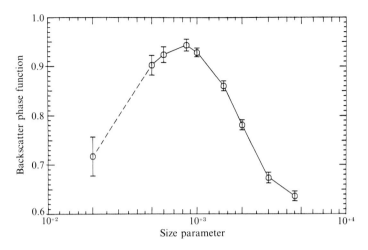

Fig. 11.3. Normalized backscatter phase function averaged over $\Delta_3\beta \approx 14$ as a function
of $\beta$ (after Shipley & Weinman 1978).

geometrical-optic value for a totally reflecting sphere, i.e., for an ideal
isotropic scatterer (van de Hulst 1957). They are given by

$$G_j(\beta,\theta) = 4 i_j(\beta,\theta)/\beta^2 \qquad (j = 1,2) \tag{11.3}$$

The 'normalized backscatter phase function' (which, when $\beta \gg 1$, is
approximately proportional to the backscatter gain factor), averaged over
the largest period $\Delta_3\beta \approx 14$, is plotted against $\beta$ in fig. 11.3. It shows a
broad peak, centered around $\beta \sim 10^3$.

(xv) *Edge origin.* Studies of the sum of the first $L$ terms of the Mie series
as a function of $L$ (Bryant & Cox 1966) indicate that the dominant
contributions to the backscattered intensity arise from the *edge strip*
(8.3)–(8.4). This is confirmed by near-field observations, in which light
arising from the droplet circumference can be distinguished from the axial
spot produced by geometrical reflection and in which incident light can be
selectively focused along the axis or on the droplet edge (Fahlen &
Bryant 1968, Saunders 1970, Ashkin & Dziedzic 1981).


## 11.2   Cross-polarization  and  axial  focusing

For $\theta = \pi$ in the Mie series (5.6), one has

$$p_l(-1) = -t_l(-1) = (-)^l\left(l + \tfrac{1}{2}\right) \qquad (11.4)$$

so that (5.6) becomes

$$S_1(\beta, \pi) = S^M(\beta) + S^E(\beta) = -S_2(\beta, \pi) \qquad (11.5)$$

where $S^M$ denotes the contribution from magnetic multipoles [first sum in (5.6)] and $S^E$ is the electric multipole contribution [second sum in (5.6)].

For nonparaxial scattering angles, the perpendicular-polarized intensity $S_1$ is dominated by the magnetic multipole contributions and the parallel-polarized one $S_2$ is dominated by the electric multipole contributions. However, for observation along the axis (forward as well as backward), the scattering plane is undefined: a parallel component viewed from one plane appears as a perpendicular component when viewed from an azimuthal plane at right angles to the first one. This simulates an interference between the two polarizations, as in (11.4), and the polarized intensities coincide. This effect has been called *cross-polarization* by van de Hulst (1947, 1957).

The effect persists in a small neighborhood of the backward direction, determined by van de Hulst's *axial focusing effect*, already mentioned in Sec. 4.2. Let us assume, as will turn out to be true, that the dominant contributions to the glory arise from a small angular momentum range, centered around

$$\Lambda \equiv kb \gg 1 \qquad (11.6)$$

The angular functions within this range in (5.6) are slowly-varying in the angular momentum, so that they can be factored out, and we can then employ the asymptotic expansions (Khare & Nussenzveig 1977b)

$$t_{\lambda-\frac{1}{2}}(\cos\varepsilon) \approx 2\left(\frac{\varepsilon}{\sin\varepsilon}\right)^{\frac{1}{2}} \lambda\, J_1'(\lambda\varepsilon)$$

$$p_{\lambda-\frac{1}{2}}(\cos\varepsilon) \approx 2\left(\frac{\varepsilon}{\sin\varepsilon}\right)^{\frac{1}{2}} \frac{J_1(\lambda\varepsilon)}{\varepsilon} \qquad (11.7)$$

where $J_1$ is Bessel's function of order one,

$$\theta = \pi - \varepsilon, \quad 0 \leq \varepsilon \ll 1 \qquad (11.8)$$

and we have represented the amplitudes in terms of

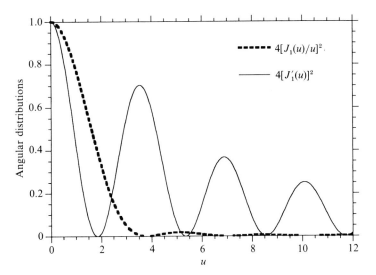

Fig. 11.4. Angular distributions for glory components.

$$t_l(-\cos\theta) = (-)^l t_l(\cos\theta) \quad , \quad p_l(-\cos\theta) = (-)^{l+1} p_l(\cos\theta) \quad (11.9)$$

For $|\lambda\varepsilon| \gg 1$, it follows from (11.7) that

$$\left|p_{\lambda-\frac{1}{2}}(\cos\varepsilon)/t_{\lambda-\frac{1}{2}}(\cos\varepsilon)\right| = O\left(|\lambda\varepsilon|^{-1}\right) \quad (11.10)$$

which leads to the dominance of magnetic terms in $S_1$ and electric ones in $S_2$. With the above assumptions, the amplitudes may be approximated by

$$S_1(\beta, \pi-\varepsilon) \approx 2S^M(\beta)J_1'(\Lambda\varepsilon) + 2S^E(\beta)\frac{J_1(\Lambda\varepsilon)}{\Lambda\varepsilon}$$
$$S_2(\beta, \pi-\varepsilon) \approx -2S^E(\beta)J_1'(\Lambda\varepsilon) - 2S^M(\beta)\frac{J_1(\Lambda\varepsilon)}{\Lambda\varepsilon} \quad (11.11)$$

where the coefficients follow from the assumption that, for $\varepsilon = 0$, the results must reduce to (11.5).

The squares of the angular factors in (11.11) are plotted in fig. 11.4. We see that the cross-polarization effect persists over the range $\Lambda\varepsilon \leq O(1)$, but that it disappears for $\Lambda\varepsilon \gg 1$.

The relationship between these results and axial focusing appears when we note that these angular factors are a vectorial generalization of

(4.1). The dominance assumption (11.6), by the localization principle, is equivalent to the assumption that the paraxial angular distribution arises from a ring source of radius $b$. The radiation from such a source may be obtained by applying a vectorial (electromagnetic) generalization of Huygens' principle. Just as (2.1) leads to (4.1) for scalar waves, the vector ring source radiation leads to (11.11) (van de Hulst 1947, 1957).

## 11.3  Geometrical-optic and van de Hulst terms

The first contributions we consider in the CAM theory of the glory are those from the lowest-order Debye terms, namely, those discussed in Chapters 9 and 10, that suffice to account for most of the observed scattering away from near-backward directions. For simplicity, we take $\theta = \pi$ and restrict our consideration to refractive indices in the range (10.1); for the meteorological glory, we have to do with water droplets.

*Geometrical-optic (WKB) contributions*

The leading contributions according to geometrical optics arise from *axial rays*: the directly reflected one and the one which undergoes backscattering after one internal reflection, associated with the third Debye term. Axial-ray contributions originate from saddle points at $\lambda = 0$, so that (11.6) does not hold; consequently, such contributions are not enhanced by axial focusing (they arise from virtual focal points rather than virtual ring sources).

The direct-reflection axial-ray contribution corresponds to (9.15) at $\theta = \pi$:

$$S_{j0}^{\text{WKB}}(\beta, \pi) \approx (-)^{j+1} \frac{i\beta}{2} \left( \frac{N-1}{N+1} \right) \left[ 1 + O(\beta^{-1}) \right] \exp(-2i\beta) \qquad (11.12)$$

The axial-ray contribution from the third Debye term (Sec. 10.1) is given by (Khare 1975, 1982)

$$S_{j2}^{\text{WKB}}(\beta, \pi) \approx (-)^{j+1} i\beta \frac{2N^2(N-1)}{(2-N)(N+1)^3} \left[ 1 + O(\beta^{-1}) \right]$$
$$\times \exp[2i(2N-1)\beta] \qquad (11.13)$$

Axial-ray contributions from higher-order Debye terms contain powers of $[(N-1)/(N+1)]^2 \approx 0.02$ for $N \approx 1.33$, so that they can be neglected (this is an illustration of the fast convergence of the Debye expansion for saddle-point contributions).

Neglecting these higher-order axial-ray contributions, the geometrical-optic result for the backscattered intensity is the sum of the squares of the absolute values of (11.12) and (11.13). It differs from the WKB result, where the phases are taken into account: one first adds up the amplitudes (11.12) and (11.13), then takes the absolute value squared.

In the WKB result, the analogue for the glory of Young's interference theory of the rainbow, there is a sinusoidal interference term, with period $\pi/(2N)$ and amplitude comparable to the geometrical-optic intensity. Averaging over the interference oscillation, one is led to the geometrical-optic result, plotted in fig. 5.2. As illustrated in this figure, the backscattered intensity predicted by geometrical optics (as well as by the WKB approximation) is typically at least one order of magnitude below that required to account for the background [feature (vii), Sec. 11.1].

We conclude that *the scattered intensity in the glory region is dominated by diffraction effects*.

### *The van de Hulst term*

Complex angular momentum theory enables us to compute the surface-wave contribution to the third Debye term, associated with diffracted rays of the type illustrated in fig. 4.3, which, according to van de Hulst's conjecture (van de Hulst 1957), would yield the leading contribution to the glory.

For this purpose, according to Sec. 10.1, one must evaluate the residues of the third Debye integrand in the Poisson representation, corresponding to $p = 2$ in (9.10), at the Regge-Debye poles $\lambda_n^{(j)}$ shown in fig. 9.4, which are triple poles for this term. The evaluation has been performed for the scattering of scalar waves (Nussenzveig 1969b), as well as for Mie scattering (Khare 1975), employing a transitional asymptotic expansion of the cylindrical functions due to Schöbe (1954). The Mie results are of the form

$$S_{12}^{sw}(\beta,\pi) = -S_{22}^{sw}(\beta,\pi) = \beta^{4/3}\exp(4iM\beta)$$

$$\times\sum_{j=1}^{2}\left(e_j\right)^{-2}\sum_{n}c_{nj}\exp\left(i\lambda_n^{(j)}\zeta_2\right) \tag{11.14}$$

where the superscript sw stands for 'surface wave', $e_j$ is defined in (5.9), and

$$M \equiv \left(N^2 - 1\right)^{1/2} \tag{11.15}$$

$$\zeta_2 \equiv \pi - \theta_L = \pi - 4\cos^{-1}\left(1/N\right) \tag{11.16}$$

In (11.14), the phase factor $\exp(4iM\beta)$ arises from the two shortcuts [AB and CD in fig. 4.3(b)] taken by the diffracted rays through the sphere, and $\zeta_2$ is the 'missing angle' for edge-incident rays to be backscattered after taking these two shortcuts [the angle spanned by the arc TA + BC + DT' in fig. 4.3(b)], representing the ~15° gap that has to be bridged by the surface waves; the angle $\theta_L$ in (11.16) is illustrated in fig. 10.1(a).

As indicated in (11.5), both polarizations contribute to the amplitudes; the contribution from polarization 2 (electric) is enhanced by a factor $(e_2/e_1)^{-2} = N^4$, which is ~ 3 for water. The numerical coefficients $c_{nj}$ in (11.14) are given by expansions into powers of $\gamma$. The lowest-order term is proportional to $\zeta_2$, and is therefore small ($\zeta_2$ is ~ 0.26 for water), so that the first correction, of order $\gamma^2$, is important, especially for low $\beta$.

As is illustrated by (8.15) and (8.18), the typical surface-wave amplitudes are $O(\beta^{-1/6})$ times smaller than WKB contributions (apart from their exponential decay factors). However, we see that (11.14) contains a factor $O(\beta^{1/3})$ times *larger* than the WKB terms (11.12) and (11.13). The difference arises from the axial focusing enhancement factor, which is of order $\beta^{1/2}$ for the surface waves, since they originate from the edge, with $\Lambda = \beta$ in (11.6).

One must still take into account the *surface-wave damping factor*

$$\exp\left(-\zeta_2 \operatorname{Im} \lambda_n^{(j)}\right) \tag{11.17}$$

for each Regge–Debye pole contribution in (11.14). Since the Regge–Debye poles are relatively close to the hard-sphere Regge poles (8.16), the damping grows rapidly with $n$, so that it is sufficient to keep the contributions from the first few poles in (11.14). As a rough order-of-magnitude estimate,

$$S_{j2}^{sw}(\beta, \pi) \sim c\beta^{4/3} \exp\left(-d\beta^{1/3}\right) \tag{11.18}$$

where $c$ and $d$ are of order unity.

For $\beta \sim 10^2$, it was verified in scalar scattering (Nussenzveig 1969b) that the van de Hulst term is dominant over the geometrical-optic

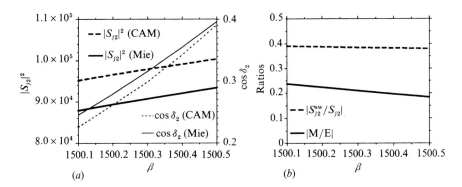

Fig. 11.5. *(a)* Comparisons between CAM and Mie results for third Debye term at $\theta = \pi$ for $N = 1.330\,07$, near $\beta = 1500$: squares of magnitudes (left scale) and cosine of phase difference between electric and magnetic contributions (right scale); *(b)* Ratios: magnitude of surface-wave contribution to total term; and magnetic to electric contribution.

ones, and that its order of magnitude agrees with that of the backscattering amplitude; however, additional contributions were found to be required to fully account for the glory features.

Numerical Mie results for the third Debye term for $N = 1.330\,07$ and $\beta$ around 1500 (Khare 1975, 1982) are plotted in fig. 11.5. In fig. 11.5*(a)*, the Mie result for the square of the magnitude of this term is compared with the corresponding CAM result, obtained by adding the WKB result (11.13) to the surface-wave contribution (11.14). The agreement is to better than 10%, the error arising mainly from the use of transitional asymptotic approximations in the computation of (11.14). We also compare in this figure (on a different scale) the cosine of the phase difference $\delta_2$ between electric and magnetic contributions to the third Debye term, for which the agreement between CAM and the Mie result is somewhat more accurate.

In fig. 11.5*(b)*, the ratio of the magnitude of the surface-wave contribution to the third Debye term to that of the total term is plotted; we see that even at this large value of $\beta$ the surface-wave amplitude is about 40% of the total, in spite of the exponential decay in (11.18). For $\beta$ of order $10^2$, as was found in the scalar case, the van de Hulst surface-wave term is the leading contribution to the third Debye term. Also plotted is the ratio of the amplitudes of the magnetic and electric contributions to the third Debye term, which is seen to be of the order of 0.2, so that the electric polarization is enhanced by a factor of about 5, in reasonable agreement with the rough estimate given above.

Thus, CAM theory does allow us to evaluate the van de Hulst

surface-wave term. Unfortunately, however, when we compare fig. 11.5(a) with the total backscattered intensity in this range, we find that the total intensity is at least one order of magnitude larger than the contribution from the third Debye term. Therefore, one cannot account for the glory by considering only the first three terms of the Debye expansion: *higher-order Debye terms must be taken into account.*

## 11.4   Orbiting and leading higher-order terms

Which higher-order Debye terms may be expected to yield a significant contribution to backscattering? The contributions from these terms tend to be strongly damped by multiple internal reflection (cf. Sec. 9.4). Unless the angle of incidence is very steep, the low value of the Fresnel reflectivities renders them negligible.

For example, the lowest-order backward glory ray (backscattered non-axial ray) for water has an angle of incidence of about 34° and four internal reflections ($p = 5$), yielding an internal reflection damping factor $\sim 10^{-3}$, so that, even with an axial focusing enhancement factor of order $\beta^{1/2}$, this contribution remains negligible within the range of size parameters of interest.

In order to minimize internal reflection damping, we must go to the *edge strip* (8.3)–(8.4). According to geometrical optics, the internal reflectivity associated with edge incidence would be 1 (total reflection). However, as we have seen in Sec. 8.4, this is not so when we take into account the dynamical effects of the surface curvature on reflection, i.e., the penetrability of the centrifugal barrier (8.2). The internal spherical reflection coefficient in the edge strip is found (Nussenzveig 1969b, 1979) to be of the form

$$\left|R_{11}^{(j)}\right| = 1 - b_j \beta^{-1/3} \tag{11.19}$$

where $b_j$ is of order unity.

The *internal-reflection damping factor* for the $(p+1)$th Debye term in the edge strip is therefore, for large $p$, of the form

$$d_p^r \equiv \left|R_{11}^{(j)}\right|^p \sim \exp\!\left(-p\,b_j\beta^{-1/3}\right) \qquad (p \gg 1) \tag{11.20}$$

and the axial focusing enhancement factor is of order $\beta^{1/2}$ [since $b \approx a$ in

(11.6)]. It follows that *Debye terms of orders up to a few times $\beta^{1/3}$, that undergo a large number of internal reflections, can still yield appreciable contributions from the edge strip.*

In terms of the effective potential, these contributions arise from a neighborhood of the top T of the centrifugal barrier in fig. 9.1(*a*), so that this effect is analogous to *orbiting* (Sec. 1.1). Indeed, the corresponding paths describe several turns around the origin before emerging.

However, not all Debye terms within the allowed range contribute equally: we must still look for the leading contributors among them.

### Geometric resonances

Additional enhancement factors that counteract internal reflection damping, besides the axial focusing enhancement, arise from situations associated with the proximity to an *edge backward glory ray* or to a near-backward *higher-order rainbow.*

In order for an incident edge ray to emerge in the backward direction after taking $p$ shortcuts through the sphere, one must have [cf. figs. 2.1 and 9.5(*b*)]

$$p(\pi - 2\theta_2) \equiv p(\pi - 2\theta_c) = (2m+1)\pi \qquad (11.21)$$

where the integer $m$ denotes the number of complete turns before backscattering and $\theta_c$ is the critical angle (9.17). It follows that one must have

$$N = \left\{ \cos\left[ \frac{(2m+1)\pi}{2p} \right] \right\}^{-1} \qquad (11.22)$$

For water, in the visible, $N$ ranges approximately from 1.33 to 1.34, so that $(2m+1)/(2p)$ is between 1/5 and 1/4. The lowest value of $p$ satisfying (11.21) then is (Khare 1975, Khare & Nussenzveig 1977a) $p = 24$, for which

$$N = \left[\cos(11\pi/48)\right]^{-1} \approx 1.330\ 07 \qquad (11.23)$$

The corresponding geometrical-optic orbit for an edge-incident ray, as illustrated in fig. 11.6, is a 48-sided regular inscribed star-shaped polygon, which would be described periodically any number of times in the absence of damping. The choice of this value for $N$, while by no means essential, simplifies the discussion.

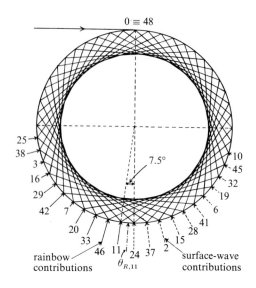

Fig. 11.6.  Closed 48-sided polygonal edge-ray orbit for $N \approx 1.330\,07$. At the vertices, the number indicates the value of $p$ and the arrow points in the direction $\zeta_p$; its length qualitatively indicates the 'naïve' ordering of surface-wave (---) and rainbow (—) contributions (neglecting reflection damping). The shift of the rainbow angle $\theta_{R,p}$ from $\zeta_p$ is indicated for $p = 11$ (after Nussenzveig 1979).

### Leading surface-wave terms

For the Debye terms whose order $p$ is indicated in fig. 11.6 next to the vertices in the fourth quadrant, the 'missing angle' $\zeta_p$ for emergence in the backscattering direction is

$$\zeta_p \equiv \pi - p\theta_t \;(\text{mod }2\pi), \quad 0 \le \zeta_p \le \theta_t \equiv 2\cos^{-1}(1/N) \qquad (11.24)$$

with $\theta_t = 11\pi/24$ for the choice (11.23); thus, for the van de Hulst term $p = 2$, $\zeta_2 = \pi/12 = 15°$ corresponds to (11.16).

As a generalization of (11.17), the residues at Regge–Debye poles for these Debye terms give rise to contributions with *surface-wave damping factors*

$$d^{\text{sw}}_{p,n} \sim \exp\!\left(-\zeta_p \,\text{Im}\,\lambda_n^{(j)}\right) \sim \exp\!\left(-v_n^{(j)} \beta^{1/3} \zeta_p\right) \qquad (11.25)$$

where $v_n^{(j)}$ increases rapidly with $n$. Taking into account only this factor, i.e., neglecting internal reflection damping, one would predict that the magnitude of surface-wave contributions to the glory from the $p$th Debye term should decrease roughly exponentially with $\zeta_p$, leading to the 'naïve' ordering of these contributions, indicated in fig. 11.6 by the decreasing length of the dashed arrows that point at increasing values of $\zeta_p$. Each of them corresponds not only to the value of $p$ shown in fig. 11.6, but also to all those that differ from it by an arbitrary multiple of the basic period $\Delta p = 48$.

The true ordering, however, is modified by the internal-reflection damping factor (11.20), which leads to exponential decay with $p$ for large $p$, suppressing all such contributions. This high-$p$-cutoff gets lower as $\beta$ decreases, so that the effect is particularly important within the range where natural glories are observed. Thus, while $p = 24$ and $p = 37$ should predominate over the van de Hulst $p = 2$ surface-wave term for large $\beta$, the reverse is true for smaller values of $\beta$.

*Leading near-backward higher-order rainbows.*

The rainbow angle of incidence for the $p$th Debye term (rainbow of order $p - 1$) is given by (10.16), so that, as was mentioned in Sec. 10.4, high-order rainbows arise from near-edge incidence. Thus, the vertices in the third quadrant of the circle in fig. 11.16, associated with *negative* values of $\zeta_p$ in (11.24), must be close to the exit points of geometrical rainbow rays for rainbows formed near the backward direction, that turn their dark side toward it.

Indeed, the angular distance from the geometrical rainbow angle $\theta_{R,p}$ to the backward direction, for large $p$, is found to be given by (Khare 1975, Nussenzveig 1979)

$$\varepsilon_{R,p} \equiv \pi - \theta_{R,p} \approx -\zeta_p - \frac{M}{p} \quad \left(\zeta_p < 0,\ p \gg 1\right) \qquad (11.26)$$

which is not too bad an approximation even for the secondary bow, $p = 3$ in fig. 11.6.

The rainbow amplitude enhancement (of order $\beta^{1/6}$) and the axial focusing enhancement (of order $\beta^{1/2}$) for such contributions are counteracted by the damping effects in the rainbow shadow, where the backward direction lies. The uniform CAM rainbow approximation is given by a generalization of (10.11) (Khare 1975, 1982), and the complex-ray *rainbow shadow damping factor* in the backward direction follows from (10.15):

$$d_p^R \sim \exp\left[-\mu_p \beta \left(\frac{\varepsilon_{R,p}}{p}\right)^{3/2}\right], \quad p \gg 1, \ \varepsilon_{R,p} \geq \Delta\theta_{R,p} = O(p\beta^{-2/3}) \quad (11.27)$$

where $\mu_p$ is of order unity.

The last inequality in (11.27) amounts to requiring that the backward direction fall within the rainbow shadow, outside of the main rainbow peak, the width $\Delta\theta_{R,p}$ of which is of order $p\gamma^2$, $p$ times larger than the angular width of the primary bow region [cf. Sec. 10.4, (vi)].

The 'naïve' ordering of rainbow shadow contributions to the glory, indicated by the length of the solid arrows in fig. 11.6, classifies them in order of decreasing importance by increasing values of $\varepsilon_{R,p}/p$. The leading contribution arises from the *tenth-order rainbow*, for which one has $\varepsilon_{R,11} \approx 0.05$ rad $\approx 3°$, so that it is very close to the backward direction.

Again, when the internal-reflection damping factor (11.20) is taken into account, the ordering is modified, particularly for the lower values of $\beta$. Thus, while $p = 33$ should dominate over $p = 46$ for low $\beta$, the reverse should be true for large $\beta$.

The order of magnitude of higher-order rainbow contributions to the glory also depends on whether the backward direction lies deep within the rainbow shadow, so that the damping factor (11.27) is fully effective, or it still lies within the main rainbow peak, thereby yielding a larger contribution. For a given $p$, the latter situation holds for low $\beta$ and the former one for high $\beta$, as is to be expected, since the width $\Delta\theta_{R,p}$ of the main rainbow peak is proportional to $p\gamma^2$ [cf. (11.27)]. The boundary between high and low $\beta$ lies at a size parameter such that the exponent in (11.27) becomes of order unity. Thus, for $p = 11$, the transition occurs for $\beta$ of the order of $10^3$.

*Verification*

Contributions from various Debye terms $p \geq 1$ to $|S^M(\beta)|^2$ and to $|S^E(\beta)|^2$ in (11.5), computed by numerical summation of the corresponding partial-wave series, for $N = 1.330\ 07$ and for $\beta = 150, 500$ and $1500$, are plotted in fig. 11.7 (Khare & Nussenzveig 1977b). Different Debye orders interfere in $|S_j(\beta,\pi)|^2$, but the interference terms tend to be wiped out by size averaging, so that the relative contributions to the meteorological glory can be estimated from this plot.

The results are in good agreement with those expected from the above discussion, both regarding the leading $p$ values and their relative ordering. Even though the contributions plotted range over three orders of magnitude, values of $p$ not included in fig. 11.6 appear only for the lowest

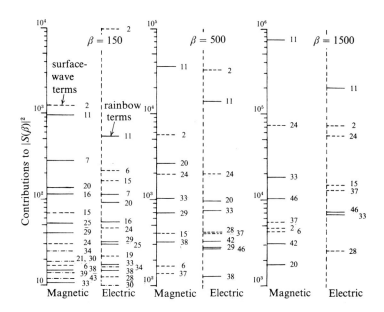

Fig. 11.7. Contributions to $|S^M(\beta)|^2$ and $|S^E(\beta)|^2$ from various Debye orders $p \geq 1$ (the values of $p$ are indicated) for $N = 1.33007$ and $\beta = 150$, 500 and 1500. $---$ Surface-wave terms; $\longrightarrow$ Rainbow terms. For $\beta = 150$, there appear some terms ($\longrightarrow\cdot\longrightarrow$) not present in fig. 11.6 (after Khare & Nussenzveig 1977b).

$\beta$, and the ordering is consistent with the expectations.

Thus, for $\beta = 150$, close to the estimated average size parameter in natural glory observations, the leading contribution arises from the van de Hulst $p = 2$ surface-wave term, followed by the tenth-order rainbow term $p = 11$ and by the next low-order rainbow ($p = 7$) and surface-wave ($p = 6$) contributions. We see that, at this low value of $\beta$, the naïve ordering of fig. 11.6 is strongly affected by internal reflection damping, which favors lower $p$ values over higher ones; for the same reason, some Debye terms not included in fig. 11.6 give (small) contributions.

For $\beta = 500$, the contributions from $p = 2$ and $p = 11$ are already comparable, followed by the $p = 20$ rainbow term and by the $p = 24$ backward glory ray.

Finally, for $\beta = 1500$, the $p = 11$ rainbow term has already become dominant, and the $p = 24$ backward glory term also dominates over $p = 2$ for the magnetic contribution. Internal-reflection damping is less significant for this higher value of $\beta$, so that the ordering is closer to the naïve one shown in fig. 11.6.

## 11.5   CAM theory of the glory

The agreement between the results in fig. 11.7 and CAM predictions
about the classification of leading contributions indicates that the physical
interpretation in terms of surface waves and higher-order rainbows must
be valid. However, for a quantitative comparison, CAM approximations to
these higher-order Debye terms are required.

### *Evaluation of dominant terms*

The evaluation of the $p = 2$ van de Hulst contribution has already been
discussed in Sec. 11.3, and we have seen (fig. 11.5) that the CAM
transitional approximation agrees with the Mie result to better than 10%
around $\beta = 1500$; the agreement should be improved by using uniform
approximations. The same procedure may be employed for the evaluation
of higher-order surface-wave contributions For $\beta \lesssim 10^2$, the van de Hulst
surface-wave contribution (11.14) is dominant.

The CAM evaluation of the $p = 11$ rainbow term (Khare 1975,
1982; Khare & Nussenzveig 1977b) leads to an expression similar to
(10.11), except for the additional enhancement due to axial focusing.
Higher-order CFU coefficients give significant contributions, so that the
Airy theory fails for *both* polarizations, as is expected for higher-order
rainbows (Sec. 10.4).

Numerical Mie results for the $p = 11$ Debye term at $\theta = \pi$, for
$N = 1.330\ 07$, near $\beta = 1500$, are compared with the CAM rainbow
approximation in fig. 11.8. We see that the agreement is good both for
$|S_{j11}|^2$ and for the phase difference $\delta_{11}$ between the electric and magnetic
contributions to this term. In contrast with surface-wave terms [cf. fig.
11.5(b)], the magnetic contribution is dominant, as is characteristic for a
rainbow term. The agreement found is a good test of CAM theory for a
higher-order rainbow.

The $p = 24$ glory-ray contribution at $\theta = \pi$ lies within a Fock-type
region. A transitional approximation to this term has also been compared
with the corresponding Mie results around $\beta = 1500$ (Khare 1975, Khare
& Nussenzveig 1977b), showing good agreement in both amplitude and
phase.

### *Numerical comparisons*

As a first approximation to $S^M(\beta)$, $S^E(\beta)$, one can take the sum of the
leading and next-to-leading contributions in fig. 11.7. For $S^M$, these are
$p = 2$ and $p = 11$ near $\beta = 150$ and (in reverse order) $\beta = 500$, and $p = 11$

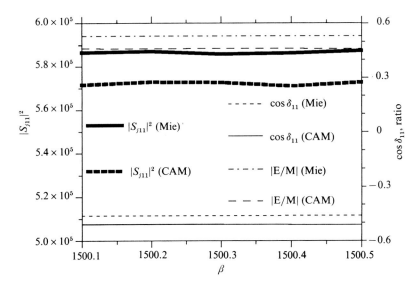

Fig. 11.8. Comparison between Mie results for $p = 11$ Debye term at $\theta = \pi$, $N = 1.330\ 07$, near $\beta = 1500$, and corresponding CAM rainbow term: squares of magnitudes (left scale) and cosine of phase difference and ratio of magnitudes between electric and magnetic contributions (right scale).

and $p = 24$ near $\beta = 1500$. In fig. 11.9, the exact results around these values of $\beta$, within an interval containing a quasiperiod of the glory pattern (curves in full line), are compared with the contribution from the leading pair of Debye terms (dash-and-dot curves). The direct-reflection term $p = 0$ is omitted in these curves.

We see that a sizable fraction of the relatively slowly-varying 'background' oscillation, with two to three humps per quasiperiod, is already accounted for by the sum of these two terms (the leading surface-wave term and the leading rainbow term).

Also shown in fig. 11.9 (curves in dashed line) is the result of including the contributions from all Debye terms that appear in fig. 11.7. This yields a close approximation to the exact results, though failing to reproduce the sharp spikes associated with the ripple fluctuations.

The inset in fig. 11.9(c) shows that a fairly sharp spike can be approximated by including, besides the terms that appear in fig. 11.7, those obtained from them by summation over the orbiting period $\Delta p = 48$. Thus, such a spike can be interpreted as a geometric resonance effect arising from the existence of quasiperiodic orbits.

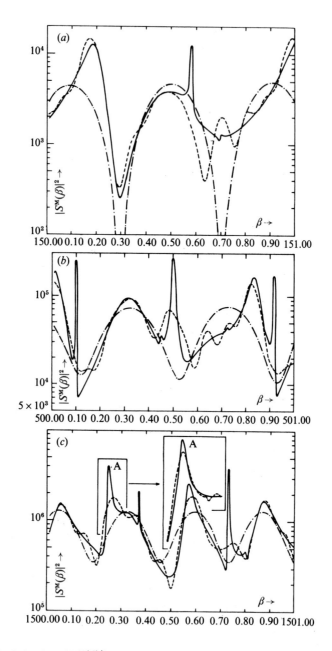

Fig. 11.9. Behavior of $|S^M(\beta)|^2$ for $N = 1.330\,07$: ——— exact; —·— leading pair of Debye terms; - - - - all Debye terms in fig. 11.7 (no summation over $\Delta p = 48$). *(a)* near $\beta = 150$; *(b)* near $\beta = 500$; *(c)* near $\beta = 1500$. The inset shows an amplification (———) of the spike at A, compared with the contribution from all Debye terms in fig. 11.7 summed over $\Delta p = 48$ (from Khare & Nussenzveig 1977a).

The need to invoke many Debye terms to account for sharp spikes is an indication that the Debye expansion is no longer appropriate to deal with them. As will be seen in Chapter 14, sharp spikes, including ripple fluctuations too narrow to be resolved in fig. 11.9, should be interpreted in terms of resonances in the total scattering amplitudes, rather than in terms of individual Debye contributions.

## 11.6  Explanation of the glory features

We now apply the results from CAM theory to discuss the explanation of the glory features listed in Sec. 11.1.

### *Variability and quasichaotic features*

The CAM theory of ripple fluctuations [feature (viii)] will be discussed in Chapter 14. The role of very sharp ripple fluctuations in natural glories is negligible (in contrast with broader ones, as is explained below). Indeed, because of their 'sensitive dependence' on the parameters $\beta$ and $N$, sharp fluctuations are washed out by size and material dispersion, as well as by the presence of absorption [feature (xi)], in view of their association with very long optical paths within the droplets. They may play a role in the observed variability of the glory [feature (i)], as well as in the quasichaotic character of the amplitude spectrum [feature (xiii)]. The occurrence of 'geometric resonances', i.e., of closed polygonal orbits with not too large Debye order $p$ (large $p$ values are quenched by internal reflection damping) also depends sensitively on $N$.

What is basically required from a theory of the meteorological glory is the explanation of the relatively slowly-varying *background* on which the sharp ripple fluctuations are superimposed.

### *Edge origin*

As illustrated by fig. 11.9, most of the background is accounted for by the sum of the leading pair of Debye contributions, a surface-wave term (including a backward glory term as a limiting case) and a rainbow term. These terms, as well as additional Debye corrections, arise from the edge strip [feature (xv)]. The broader ripple fluctuations can contribute very substantially to the glory background, but these, as will be seen in Chapter 14, also arise from the edge strip.

### *Intensity enhancement*

The selection of the leading pair of Debye contributions, as well as their ordering, is strongly size-dependent, due to the effect of internal-reflection damping (which favors lower $p$, but becomes less effective as $\beta$ increases) on the naïve ordering.

Thus, for $\beta$ of order $10^2$, the range associated with natural glories, the leading background Debye term is the van de Hulst $p = 2$ term, dominated by surface waves (though including also the leading geometrical-optic axial-ray contribution). Because of the axial focusing effect, the surface-wave contribution, in this range of $\beta$, gets enhanced sufficiently to counteract surface-wave damping, thus contributing to the explanation of the glory intensity enhancement [feature (vii)].

For $\beta$ of order $10^3$, internal-reflection damping for $p = 11$ has decreased enough so that this term, dominated by the tenth-order rainbow, becomes the leading Debye contributor to the background. The intensity is now enhanced by the combined effects of the rainbow and axial focusing enhancements. The next-to-leading contribution no longer arises from the van de Hulst term, but rather from the $p = 24$ backward glory term.

### Quasiperiodicity

The superposition of the dominant Debye terms represented in fig. 11.7 gives rise to an interference pattern governed by their relative phase differences. Each shortcut through the sphere in fig. 11.6 contributes to the phase a term

$$2Nka\cos\theta_c = 2M\beta$$

where $\theta_c$ is the critical angle and $M = (N^2 - 1)^{1/2}$ [cf. (11.15)]. Thus, with $N = 1.330\ 07$, a change in $\beta$ by

$$\Delta\beta = 5\pi/(22M) \approx 0.814 \qquad (11.28)$$

changes the phase of the dominant Debye terms in fig. 11.7 either by $\pi$ or by an amount close to $\pi$ (mod $2\pi$). This explains the periodicity of the background and agrees with the observed quasiperiod [feature (ix)]. As will be seen in Chapter 14, the ripple fluctuations have a nearly identical quasiperiod.

### Background

As illustrated in fig. 11.9, a substantial portion of the background [feature (x)] is accounted for by the contributions from the dominant surface-wave term and the dominant rainbow term. For values of $\beta$ up to a few hundred, we obtain a reasonable first approximation to the background by also including the leading geometrical-optic axial-ray contributions (11.12) and (11.13) (other significant contributions are discussed in Chapter 14):

$$S_j \approx S_{j2}^{\text{sw}} + S_{j11}^{\text{R}} + S_{j2}^{\text{WKB}} + S_{j0}^{\text{WKB}} \tag{11.29}$$

where the van de Hulst surface-wave term $S_{j2}^{\text{sw}}$ is given by (11.14) and the tenth-order rainbow term $S_{j11}^{\text{R}}$ is given by an expression analogous to (10.11). The terms in (11.29) are in order of decreasing importance.

### Other quasiperiods

In $|S_j|^2$, interference among the terms in (11.29) gives rise to oscillations with amplitudes that are slowly-varying within a basic quasiperiod (11.28). The corresponding quasiperiods (Nussenzveig 1969b, 1979) are readily computed. One finds $\Delta_1\beta \approx 0.41$ (interference between $S_{j2}^{\text{sw}}$ and $S_{j11}^{\text{R}}$), $\Delta_2\beta \approx 1.1$ (interference between $S_{j2}^{\text{sw}}$ and $S_{j0}^{\text{WKB}}$) and $\Delta_3\beta \approx 14$ (interference between $S_{j2}^{\text{sw}}$ and $S_{j2}^{\text{WKB}}$). Thus, feature (xii) is explained.

### Average phase function

To discuss the average phase function plotted in fig. 11.3, we must, according to (11.3), average $|S_j|^2/\beta^2$ over the largest period $\Delta_3\beta \approx 14$. In the approximation (11.29), the average erases the interference terms, yielding the sum of the squared moduli of the four contributions included in this expression. The last two terms yield the geometrical-optic contribution, which is just a small constant background term (cf. fig. 5.2).

The van de Hulst surface-wave contribution to the average normalized phase function, according to (11.18), is roughly of order

$$|c|^2 \beta^{2/3} \exp\left(-2d\beta^{1/3}\right) \quad \left[(c,d) = O(1)\right]$$

which is dominant until $\beta$ reaches a few hundred. Thereafter, its relative contribution decreases and is surpassed by the tenth-order rainbow contribution, which contains a factor $\beta^{4/3}$ and terms in $\text{Ai}^2$ and $\text{Ai}'^2$, with relatively small coefficients [cf. (10.11)].

Because of the width increase for this higher-order rainbow, the backward direction falls within the main peak up to $\beta \sim 10^3$; thereafter, it moves progressively deeper into the rainbow shadow as $\beta$ increases, so that the complex-ray damping (11.27) becomes effective. This explains the peaking of the average phase function around $\beta \sim 10^3$ and its subsequent decrease [feature (xiv) and fig. 11.3].

*Angular width*

To discuss the angular distribution and polarization features, we employ the approximation (11.11), valid for near-backward scattering, where we may now set [feature (xv)]

$$\Lambda = \beta$$

We have, therefore,

$$S_1(\beta,\theta) \approx 2S^M(\beta)J_1'(u) + 2S^E(\beta)\frac{J_1(u)}{u},$$

$$u \equiv \beta(\pi - \theta) \le O(1) \tag{11.30}$$

and $-S_2(\beta,\theta)$ is obtained by interchanging $S^M$ and $S^E$. Expressions of this form, with unknown coefficients $c_1$ and $c_2$ respectively corresponding to $2S^M$ and $2S^E$, were first proposed by van de Hulst (1947, 1957). An immediate consequence of (11.30) (cf. fig. 11.4) is that the angular width of the glory is of order $\beta^{-1}$ [feature (iv)].

For $u \gg 1$, we have

$$J_1(u)/u = O(u^{-3/2}), \quad J_1'(u) = O(u^{-1/2})$$

so that the first term of (11.29) falls off much more slowly than the second one, i.e., than an Airy diffraction pattern [feature (ii)]. Thus, for $\pi - \theta \gg \beta^{-1}$, $S_1$ is dominated by $S^M$ and $S_2$ by $S^E$, as expected (no cross-polarization outside of the glory region).

*Angular distribution and polarization*

Within the glory region, for natural incident light, the angular distribution and degree of linear polarization are respectively given by (11.1) and (5.5), so that they depend on the relative magnitude and phase of $S^M$ and $S^E$ in (11.30).

As was pointed out following (11.16), $|S^E|$ is enhanced by a factor $N^4 \sim 3$ as compared with $|S^M|$ in the Regge-Debye residues, so that, in the surface-wave contribution to the polarized intensities, the electric polarization dominates over the magnetic one by about one order of magnitude (cf. fig. 11.7). Thus, for natural glories ($\beta \lesssim 10^2$), dominated by surface waves, the outer glory rings tend to be predominantly parallel-polarized. By inspecting the graphs in fig. 11.4, we see (van de Hulst

1957) that the dominance of $|S^E|$ also implies that the first dark ring tends to be rather hazy [features (iii), (v)].

On the other hand, as happens in the primary bow and for similar reasons (Sec. 10.4), perpendicular polarization is dominant in the tenth-order rainbow (fig. 11.7). Consequently, for $\beta \gtrsim 10^3$, when the rainbow contribution is the leading one, the outer glory rings tend to be predominantly perpendicular-polarized. The angular distribution and the polarization undergo considerable variations within the range $10^2 \lesssim \beta \lesssim 10^3$, reflecting the changing rank of various Debye contributions among the leading terms [feature (vi)].

### Conclusion

We see that CAM theory allows us to explain all the features described in Sec. 11.1. The glory arises from rays incident within the edge strip, i.e., near the top of the centrifugal barrier in fig. 9.1.

The edge strip plays the role of a virtual ring source, leading to the backward glory intensity enhancement by van de Hulst's *axial focusing* mechanism. Within the glory region, *cross-polarization* effects must be taken into account.

After penetrating within a droplet, rays incident beyond the edge propagate beyond or at the critical angle, so that they are almost totally reflected internally. The small departure from total reflection is a diffraction effect due to the surface curvature (tunneling across the centrifugal barrier), so that it decreases as the size parameter increases. Thus, many Debye terms contribute: a ray undergoes a large number of internal reflections before reemerging, an effect analogous to *orbiting*.

Among higher-order Debye contributions emerging at or close to $\theta = \pi$, those from surface waves are damped by tangential radiation as they travel along the surface, so that, in the high-$\beta$ limit, their ordering goes inversely with the 'missing angle'. In particular, for water droplets, one is close to a backward glory ray at $p = 24$. However, for smaller $\beta$, in the range where natural glories are observed, large-$p$ terms are removed by internal-reflection damping (tunneling), so that van de Hulst's $p = 2$ *surface-wave term* is the leading contributor to the meteorological glory.

Higher-order near-backward rainbow enhancements are also selected among Debye contributions, with predominance of the *tenth-order rainbow*, formed very close to $\theta = \pi$. For $\beta$ beyond a few hundred, this becomes the leading term, but the width of the main rainbow peak decreases as $\beta$ increases, and complex-ray damping in the rainbow shadow then reduces the backward gain factor.

Quasiperiodic features of the glory pattern are related with interference oscillations among the Debye components, associated with various numbers of shortcuts through the droplets. The sensitive dependence of lower-order *geometric resonances*, corresponding to closed or nearly closed quasiperiodic orbits, on variations of the droplet size and material parameters, which also holds for resonances in the total amplitude (ripple fluctuations), accounts for the observed variability and quasichaotic features of the glory.

*Competition among various types of gain* (axial focusing, rainbow enhancements) *and losses* (surface-wave radiation damping, complex-ray rainbow damping, internal-reflection damping) leads to variations in the nature and ranking of leading contributions, evolving with the size parameter, and to corresponding changes in angular distribution and polarization.

It will be seen in Chapter 14 that the slowest ripple fluctuations also contribute strongly to the glory. They correspond to *resonances in the total scattering amplitudes*, that cannot be represented as the sum of a finite number of terms of the Debye expansion: an infinite number of branches of the deflection function contribute, as is characteristic of orbiting (Sec. 1.1). These resonances arise from complex Regge poles located in the neighborhood of $\lambda = \beta$ (region 1 in fig. 9.2). They are also connected with tunneling around the edge strip.

In conclusion, therefore, we see that all leading contributions to the glory arise from complex critical points (poles or saddle points), so that this beautiful and impressive meteorological effect is produced almost entirely by *light tunneling* on a macroscopic scale.

# 12

# Near-critical scattering

In Chapters 9 to 11, the discussion of Mie scattering was restricted to
$N > 1$. We now consider a new type of semiclassical diffraction effect,
not included in the classification given in Chapter 1, that takes place, for
$N < 1$, near the critical angle for total reflection. It occurs, for example, in
the scattering of light by air bubbles in water.

In the geometrical-optic limit, the new effect, *near-critical
scattering*, does not correspond to a caustic singularity in the light
intensity, which remains continuous: there is a limiting singularity, but only
in the *gradient* of the intensity. We refer to it as a *weak caustic*.

The penetration of light incident beyond the critical angle into the
optically rarer medium is the earliest example of tunneling in physics. It
was discovered, amazingly enough, by Newton, who observed frustrated
total reflection in his experiments on Newton's rings (see the above
quotation).

According to Newton's ideas, a light ray would follow a parabolic
path within the rarer medium before returning to the denser one. This
would produce a lateral displacement of the reflected beam. Although our
views about paths differ from Newton's, the displacement does exist: it is
known as the *Goos–Hänchen shift* (Goos & Hänchen 1947). In near-
critical scattering from a curved interface, it is modified by the dynamical
effects of curvature.

We treat the new effect by a zero-order transitional CAM
approximation to near-critical scattering (Fiedler-Ferrari, Nussenzveig &
Wiscombe 1991), that leads to new types of diffraction integrals and
already provides a good approximation to the Mie amplitudes for large
size parameters.

## 12.1   Geometrical-optic theory

The first account of near-critical scattering was given by Pulfrich (1888), who observed it in the scattering of natural light from a cloud of air bubbles in water. He described the appearance of pale colors near the critical scattering angle

$$\theta_t = \pi - 2\theta_c \qquad\qquad (12.1)$$

which corresponds to $\approx 82.8°$ for $N \approx 0.75$, the relative refractive index of air with respect to water (for a color picture, see Marston 1991).

Pulfrich tried to establish a reciprocity relation between his observation and the rainbow: the cloud of air bubbles in water would correspond to a cloud of water droplets in air, with a reciprocal relative refractive index; also, $\theta_t$ is the minimal deviation of partially reflected rays, beyond which reflection becomes total – just as the primary rainbow angle marks the minimal deviation of $p = 2$ rays for $N > 1$.

Like the rainbow, the effect is structurally stable (Sec. 10.6), surviving deformation, e.g., to cylindrical geometry. Pulfrich observed it with cylindrical 'bubbles', and color photographs taken with this geometry have been reproduced (Marston *et al.* 1983a).

The diffraction pattern in monochromatic light is also similar to the rainbow one: one finds 'supernumerary' oscillations on the 'bright' (total reflection) side, and fast decay on the 'dark' (partial reflection) side. It is also modulated by much faster 'fine structure' oscillations which, as happens with the rainbow (Sec. 10.5), arise from interference with ray paths totally unrelated with the diffraction effect (Langley & Marston 1984).

However, the analogy with the rainbow is deceptive. In the geometrical-optic limit, the rainbow angle is a caustic direction, where the scattered intensity $i(\theta)$ undergoes an infinite discontinuity, whereas $i(\theta)$ (in contrast also with a geometrical shadow boundary) remains continuous at the critical scattering angle in this limit: only its derivative with respect to angle, $di/d\theta$, goes through an infinite discontinuity at $\theta_t$. This singularity, which we call a *weak caustic*, originates from the behavior of the Fresnel reflectivities at the critical angle: as is well-known (Sommerfeld 1954), they approach total reflection with a vertical slope, which then changes discontinuously to horizontal.

The geometrical-optic theory of light scattering by an air bubble in water was developed by Davis (1955). At $\theta < \theta_t$, the step-like singularity

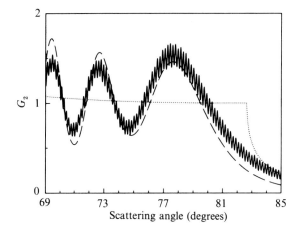

Fig. 12.1.  Gain function for polarization 2, $N = 0.75$, $\beta = 1633$. The
Mie result (solid curve) shows fine-structure oscillations.The dotted
curve is the geometrical-optic result. Also shown (dashed curve) is a
physical optics approximation (after Langley & Marston 1984)

of the Fresnel reflectivities is reproduced in the angular distribution (fig.
12.1).

How does this compare with Mie theory and with experiment?
Computed Mie results around $\theta = \theta_t$ for $\beta = 1633$ (Langley & Marston
1984) are shown in fig. 12.1. One recognizes a broad diffraction pattern
modulated by rapid fine-structure oscillations, with a main peak followed
by 'supernumeraries' on the total reflection side $\theta < \theta_t$, and a fast fall-off
on the partial reflection side $\theta > \theta_t$. Experimental results (Langley &
Marston 1984) reproduce the main features of the Mie pattern.

## 12.2  Removal of fine structure

In order to retrieve the diffraction pattern associated with near-critical
scattering, we must get rid of the obscuring effect of fine-structure
oscillations. This turns out to be much harder here than in rainbow
scattering.

In the CAM treatment (Fiedler-Ferrari *et al.* 1991), to the lowest
order of approximation, the Poisson transform of the Mie amplitudes (5.6)
in the angular domain of interest can be greatly simplified: its leading term
is given by

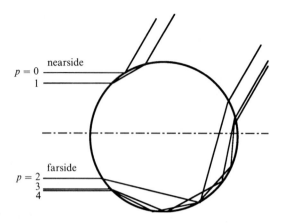

Fig. 12.2 Low-order nearside and farside paths emerging at $\theta = 60°$ ($N = 0.75$).

$$S_j(\beta,\theta) \approx -\int_0^\beta S^{(j)}(\lambda,\beta) P_{\lambda-\frac{1}{2}}(\cos\theta)\,\lambda\,d\lambda \qquad (12.2)$$

where $S^{(j)}$ is the $S$-function (9.1). In terms of the localization principle, (12.2) includes only incidence up to the edge: beyond-edge rays do not contribute significantly to the effect.

Rays incident below the critical angle get refracted into the sphere and multiply reflected inside. We can distinguish between rays traveling clockwise and counterclockwise around the sphere, by applying to (12.2) the decomposition (7.22) into running angular waves: the nearside (farside) amplitudes are obtained by substituting $P_{\lambda-\frac{1}{2}}$ by $Q_{\lambda-\frac{1}{2}}^{(1)}$ (respectively $Q_{\lambda-\frac{1}{2}}^{(2)}$). In applications to nuclear physics (Brink 1985), this is known as the *nearside–farside decomposition,* with incidence on the upper hemisphere denoted as 'nearside' (clockwise paths) and incidence on the lower hemisphere denoted as 'farside' (counterclockwise paths).

Fig. 12.2 illustrates some of the lowest-order nearside and farside paths associated with multiple internal reflections that emerge at the same scattering angle, for $\theta = 60°$ and $N = 0.75$. As one approaches $\theta = \theta_l$, the nearside paths approach critical incidence and they tend to travel internally closer and closer to the circumference, so that their relative path differences become smaller and smaller.

In contrast, the farside paths shown are still far from critical incidence: they bear no relationship to near-critical scattering effects. Because of their large path differences with respect to nearside paths, they give rise, by interference, to the rapidly-oscillating fine structure. Thus, in order to remove the perturbing effect of these oscillations, we must subtract out from the Mie amplitudes the contributions from farside paths.

These contributions can be obtained by applying the Debye expansion to the farside component of (12.2) and evaluating the resulting Debye terms in the WKB approximation (Fiedler-Ferrari *et al.* 1991). The lowest-order Debye contributions, $p = 2, 3, 4$, are represented in fig. 12.2. As may be observed in this figure, when $p$ increases for a fixed $\theta$, farside paths also tend to approach critical incidence (though much more slowly than nearside ones), leading to strong internal reflection and a relatively low rate of convergence for large $p$. This problem is less severe for polarization 2, because of an accidental proximity to Brewster's angle.

For polarization 1, this results in a complicated interference pattern, involving beats among several components with comparable amplitudes, and the number of subtractions needed before most of the fine structure is removed becomes quite large. This is illustrated by fig. 12.3, that shows the effect of 'peeling away' successive layers of fine structure from the exact polarization 1 gain factor $G_1(\beta,\theta)$ [defined by (11.3)] for $N = 0.75$ and $\beta = 10\ 000$. What remains after most of the fine structure has been eliminated is a much broader diffraction pattern (cf. also fig. 12.1), showing the features of near-critical scattering that require explanation.

## 12.3 Interference and physical optics theories

The 'interference theory' of near-critical scattering can be defined, like the corresponding theory for the rainbow (Sec. 3.2), as the result obtained by adding up the geometrical-optic amplitudes from different paths, with their phases taken into account: it is identical to the zero-order WKB approximation (primitive semiclassical approximation of Sec. 1.2).

In the vicinity of $\theta = \theta_t$, the leading nearside WKB contributions are those associated with the direct reflection and direct transmission paths shown in fig. 12.2 (for angles of incidence $\theta_1$ below the critical angle, $\theta_1 < \theta_c$) and with total reflection (for $\theta_1 > \theta_c$). They are respectively given by (9.15) and (9.20), for $\theta_1 < \theta_c$, and by (9.15) with the substitution

$$\left[N^2 - \cos^2(\theta/2)\right]^{1/2} \rightarrow i\left[\cos^2(\theta/2) - N^2\right]^{1/2}, \quad \theta < \theta_t \qquad (12.3)$$

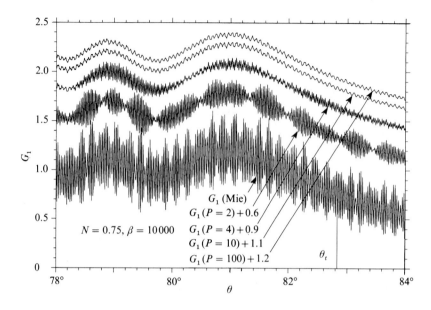

Fig. 12.3. Comparison of exact polarization 1 gain function $G_1$(Mie) for $N = 0.75$, $\beta = 10\,000$, with those where farside WKB Debye terms up to $p = P$ are subtracted out, for $P = 2, 4, 10$ and $100$. To avoid overlap, different offsets are applied to the curves. The critical scattering angle $\theta_t$ is also indicated (after Fiedler-Ferrari *et al.* 1991).

for the total reflection WKB amplitude. This is the usual substitution for the Fresnel reflection amplitudes in the domain of total reflection (Born & Wolf 1959). Since $\theta = \pi - 2\theta_1$ for direct reflection, the *domain of partial reflection*, where only the reflected path contributes, corresponds to $\theta > \theta_t$, and the *domain of total reflection*, where both reflected and transmitted paths contribute, corresponds to $\theta < \theta_t$.

Although the conditions for the applicability of the WKB approximation are violated in the vicinity of the critical scattering angle, it is useful to extrapolate it all the way to include $\theta = \theta_t$. The results for the nearside gain factor $G_1$ with $N = 0.75$, $\beta = 5000$, are compared with the subtracted Mie result in fig. 12.4.

We see that the step-like break at $\theta = \theta_t$ persists in the WKB approximation (cf. fig. 12.1). Within the domain of partial reflection $\theta > \theta_t$, the WKB result coincides with the geometrical-optic one, reproducing the rapid decay of the Fresnel reflectivity. The oscillatory behavior in the total reflection domain $\theta < \theta_t$ results from the interference between the contributions from total reflection and direct transmission. The angular scale of oscillation is set by the parameter

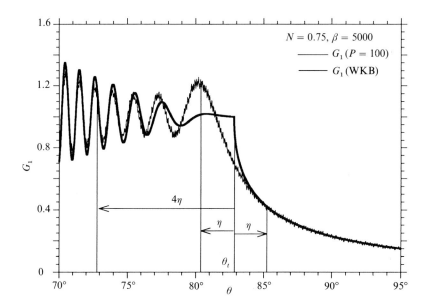

Fig. 12.4. Comparison between subtracted Mie results to order $P = 100$ and the WKB approximation: gain for polarization 1, with $N = 0.75$ and $\beta = 5000$ (after Fiedler-Ferrari *et al.* 1991).

$$\eta \equiv \sqrt{2\pi/(M'\beta)}, \quad M' \equiv \sqrt{1 - N^2} \tag{12.4}$$

which can be readily derived from the phase difference between (9.15) and (9.20) near the critical scattering angle.

The domain of applicability of the WKB approximation is predicted (Fiedler-Ferrari *et al.* 1991) to be given by $|\theta - \theta_t| \gg \eta$, and this is confirmed by fig. 12.4. On the partial reflection side, the WKB result approaches the Mie solution very rapidly, within a distance $\eta$ of the critical angle; on the total reflection side, the approach is much slower, and one needs to go an angular distance at least 4–5 times $\eta$ for the results to merge.

What is not taken into account in the WKB result is diffraction. Thus, the *angular width of the near-critical region*, where the new diffraction effects take place, is of the order of a few times $\eta$. Note that this angular width is broader than that of a rainbow, which is of order $\beta^{-2/3}$ (Sec. 10.4).

To treat the diffraction effects in the near-critical region, Marston (1979) proposed a *physical optics approximation* constructed along similar lines to the Airy theory of the rainbow (Sec. 3.2). In a Huygens–Fresnel representation, the intersection of a critically incident ray with the surface of the sphere is a stationary-phase point for critical reflection. Thus, with the origin at this point, the corresponding wavefront has a parabolic shape near the origin (the analogue of Airy's cubic wavefront).

One then applies a Kirchhoff-type approximation to the reflection amplitude along this wavefront. In view of the rapid decrease of the Fresnel reflectivities below critical incidence, they are replaced by step functions, neglecting the contribution from below-critical direct reflection. For an angle of incidence $\theta_1 = \theta_c + \varepsilon$ above the critical one but close to it $(0 \leq \varepsilon \ll 1)$, the Fresnel reflection amplitudes are pure phase factors of the form

$$R^{(j)}\left(\theta_1 = \theta_c + \varepsilon\right) \approx \exp\left(-2i\,e_j\sqrt{\frac{2N}{M'}}\varepsilon\right) \qquad (12.5)$$

corresponding to the expression within curly brackets in (9.15) for

$$\theta = \theta_t - 2\varepsilon \qquad (12.6)$$

In the physical-optics approximation, the phase factor (12.5) in the Huygens–Fresnel integral is treated as a slowly-varying one and taken at the stationary-phase point. The resulting near-critical diffraction pattern, in view of the cutoff of this integral below critical incidence and of the parabolic approximation to the wavefront, is represented by a Fresnel integral associated with the characteristic parameter (12.4).

In contrast with a Fresnel integral, however, the 'supernumerary' oscillations in the actual diffraction pattern (fig. 12.4) grow in amplitude, rather than decaying, as one gets further away from the critical scattering angle. As was discussed above, these oscillations arise from interference between total reflection and direct transmission, and their growth reflects the increasing amplitude of direct transmission as one moves away from critical incidence.

Improved versions of the physical optics approximation, taking into account the contribution from direct transmission (Marston & Kingsbury 1981), as well as adjustments to the phases of the contributing terms (Langley & Marston 1984), lead to results like those plotted in fig. 12.1 (cf. also fig. 12.8 below). While some qualitative features of the Mie results are reproduced, the quantitative agreement is poor: the physical-optics approximation does not account for near-critical scattering.

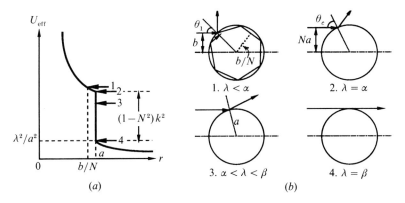

Fig. 12.5. *(a)* Effective potential for $N < 1$, showing four 'energy levels'. *(b)* Corresponding incident rays and impact parameters. 1: subcritical incidence; 2: critical incidence; 3: supracritical incidence; 4: edge incidence (after Fiedler-Ferrari *et al.* 1991).

A modified version of the physical-optics approximation (Fiedler-Ferrari *et al.* 1991), in which the contribution from below-critical direct reflection is also included, improves the agreement to some extent in the partial-reflection domain, but the errors remain considerable, with little improvement over the WKB approximation.

## 12.4   Effective potential and leading CAM terms

The effective potential for $N < 1$ [fig. 12.5*(a)*] differs from that for $N > 1$ [fig. 9.1*(a)*] by the opposite sign of the discontinuity at $r = a$, leading to a rounded potential step instead of a potential pocket. The four different values of the 'energy' $k^2$ shown in fig. 12.5*(a)* correspond to different impact parameters $b$ and angles of incidence $\theta_1$ through the localization principle:

$$b = \lambda/k = a \sin\theta_1 \tag{12.7}$$

The corresponding incident rays are represented in fig. 12.5*(b)*.

In situation 1, with $0 \leq \lambda < \alpha \equiv N\beta$, $\theta_1$ is below the critical angle, so that the incident ray is refracted into the sphere, undergoing multiple internal reflections. The radial turning point $r = b/N$ gets closer and closer to the surface as $\theta_1 \to \theta_c$. In the nearside amplitudes [cf. the discussion following (12.2)]

$$S_j^{(-)}(\beta,\theta) \approx -\int_0^\beta S^{(j)}(\lambda,\beta) Q_{\lambda-\frac{1}{2}}^{(1)}(\cos\theta) \lambda \, d\lambda \qquad (12.8)$$

the corresponding terms are obtained by applying the Debye expansion to the integral from 0 to $\alpha$.

The first (second) Debye term $p = 0$ ($p = 1$) corresponds to the partially reflected (directly transmitted) nearside rays in fig. 12.2; we refer to it as *direct reflection subcritical term* (*direct transmission subcritical term*). In the near-critical region, the contribution from direct transmission is much smaller than that from direct reflection, because of the decrease in transmissivity as total reflection is approached. For similar reasons (Fiedler-Ferrari *et al.* 1991), we can neglect $p \geq 2$ nearside Debye contributions in this region.

Situation 2 in fig. 12.5 corresponds to *critical incidence*, with impact parameter $b = Na$. The incident ray is totally reflected, but there is some *penetration by tunneling* within the sphere, corresponding to the generation of evanescent waves. The same applies to situation 3, where $\alpha < \lambda < \beta$, with the tunneling penetration depth decreasing monotonically as $\lambda$ increases. Since incident rays in this range are totally reflected, the Debye expansion should no longer be applied. Thus, the corresponding contribution, referred to as the *supracritical amplitude*, is given by the integral (12.8) extended from $\alpha$ to $\beta$, where $S^{(j)}$ is the total $S$-function (9.1).

Finally, situation 4 (fig. 12.5), where the 'energy level' lies at the bottom of the step, corresponds to edge incidence, like situation 2 in fig. 9.1(*b*). The effective potential at $r = a$ looks like a vertical wall and we expect the same physical effects found for an impenetrable sphere: the generation of external surface waves (creeping modes). The resulting contributions, as well as those from $\lambda > \beta$, are negligible in the near-critical region.

## 12.5 CAM theory of near-critical scattering

According to the discussion given in the preceding section, the leading nearside contributions in the near-critical region arise from the direct reflection and direct transmission subcritical terms and from the supracritical amplitude. Outside of the near-critical region, one can apply the WKB approximation, and both the direct reflection and supracritical contributions are dominated by the geometrical-reflection saddle point

(7.21), corresponding to (12.7) with $\theta_1 = \frac{1}{2}(\pi - \theta)$. As $\theta \to \theta_t$, this saddle point approaches $\lambda = \alpha$, and the same holds for the direct transmission geometrical saddle point, so that all leading contributions to the amplitudes in the near-critical region arise from the neighborhood of $\lambda = \alpha$ (both by the usual saddle-point arguments and because of the sharp peaking of the reflectivities at the critical angle).

In this neighborhood, the dominant optical-path term in the phase of the integrand of (12.8) can be approximated by a quadratic function, so that the reflection amplitudes (both subcritical and supracritical) have Fresnel-like phase factors in their integrands. However, the split at $\lambda = \alpha$ falls within the range [cf. (10.4)] of the geometrical-optic saddle point and the form of the integrand changes from $\lambda < \alpha$ to $\lambda > \alpha$.

*Partial reflection and transmission subcritical terms*

The WKB approximation to direct reflection, which is obtained by taking the geometrical-optic saddle-point contribution, breaks down for two different reasons:

(i) Because of the split at $\lambda = \alpha$, only part of the range of the saddle point falls within the domain of integration (in the total-reflection region $\theta < \theta_t$, only the tail end of the range). Thus, one is led to an integral with an incomplete Fresnel-like character.

(ii) The dynamical effect of the surface curvature must be taken into account in the expression for the spherical reflection coefficients, producing deviations from their limiting plane-interface Fresnel values. In a transitional asymptotic approximation, this leads to Fock-type terms in the integrands.

The zero-order transitional CAM approximation yields as a result (Fiedler-Ferrari *et al.* 1991)

$$S_{j,\mathrm{PR}}^{(-)}(\beta,\theta) \approx \frac{e^{-i\pi/4}}{2\sqrt{\pi}}\,\beta \exp\!\left(-2i\beta \sin\frac{\theta}{2}\right) \int_{(-1+i)\infty}^{X=0} \exp\!\left(-iu^2\right)$$

$$\times \left\{ \frac{1 + e^{-i\pi/6}\left(Ne_j/M'\right)\gamma'\ln'\,\mathrm{Ai}\!\left(e^{-2i\pi/3}X\right)}{1 - e^{-i\pi/6}\left(Ne_j/M'\right)\gamma'\ln'\,\mathrm{Ai}\!\left(e^{-2i\pi/3}X\right)} \right\} du \qquad (12.9)$$

where the superscript (−) stands for 'nearside' and the subscript 'PR' for 'partial reflection',

$$\gamma' \equiv (2/N\beta)^{1/3} \qquad (12.10)$$

and

$$X \equiv \gamma' \left\{ \beta \left[ \cos(\theta/2) - N \right] + \sqrt{\beta} \sin(\theta/2) \, u \right\} \tag{12.11}$$

The path of integration at infinity in (12.9) runs parallel to the steepest-descent one, so that the integrand has fast (Gaussian) decay. In the domain of partial reflection $\theta > \theta_t$, the path is taken first to the saddle point $u = 0$, then along the real axis to the upper limit, yielding a mixed steepest-descent-plus-stationary phase path analogous to the path $\Gamma'$ in fig. 8.2. The integrand combines Fresnel and Fock-type behavior.

For the directly transmitted rays (fig. 12.2), $\theta = \theta_t$ is a Fock-type shadow boundary, and the near-critical domain is a penumbra region. The direct transmission term (Fiedler-Ferrari *et al.* 1991) is given by an incomplete Fock function, similar to those of Secs. 7.3 and 8.5 but with an incomplete character arising from the cutoff at the critical angle of incidence. As was already mentioned, direct transmission contributes far less than direct reflection in the near-critical domain, but the relative contribution from this term increases as one moves away from the critical angle.

*Total reflection (supracritical) term*

The only difference between the partial-reflection (subcritical) term and the total-reflection (supracritical) one, apart from the different range of integration in the $\lambda$ plane, lies in the replacement

$$R_{22}^{(j)}(\lambda,\beta) = -\frac{\{2\beta\} - Ne_j\{2\alpha\}}{\{1\beta\} - Ne_j\{2\alpha\}} \to R^{(j)}(\lambda,\beta) \equiv -\frac{\{2\beta\} - Ne_j\{\alpha\}}{\{1\beta\} - Ne_j\{\alpha\}} \tag{12.12}$$

where the left-hand side is the spherical reflection coefficient (9.4) in the partial-reflection range and the right-hand side, contained in the total $S$-function (9.1), represents total reflection (no Debye expansion); for real $\lambda$, $R^{(j)}$ is a phase factor.

Corresponding to (12.9), one finds for this term (Fiedler-Ferrari *et al.* 1991)

$$S_{j,\mathrm{TR}}^{(-)}(\beta,\theta) \approx \frac{e^{-i\pi/4}}{2\sqrt{\pi}} \beta \exp\left(-2i\beta \sin\frac{\theta}{2}\right) \int_{X=0}^{(1-i)\infty} \exp(-iu^2)$$

$$\times \exp\left\{ 2i \tan^{-1}\left[ \frac{Ne_j}{M'} \gamma' \ln' \mathrm{Ai}(X) \right] \right\} du \tag{12.13}$$

where the subscript 'TR' stands for 'total reflection' and the path of integration at infinity is still parallel to the steepest-descent one, yielding fast numerical convergence. One now takes a hybrid steepest-descent-plus-stationary-phase path for $\theta < \theta_t$ (domain of partial reflection).

The integral (12.13) is a new type of diffraction integral, which we call a *Pearcey–Fock integral*, for reasons explained in the next section, where we discuss the physical interpretation of the CAM results.

## 12.6 Planar reflection limit and Goos–Hänchen shift

### *Planar reflection limit and Pearcey's integral*
For large values of $\beta$ in the near-critical region, the result (12.13) can be approximated, with some loss of accuracy, by (Fiedler-Ferrari 1983, Fiedler-Ferrari & Nussenzveig 1987)

$$S_{j,\mathrm{TR}}^{(-)} \approx \frac{e^{-i\pi/4}}{2\sqrt{\pi}} \beta \exp\left[-2i\beta(M'-N\varepsilon)\right]$$

$$\times \int_0^\infty \exp\left\{-i\left[\zeta^2 - x\zeta - \frac{y^{4/3}}{2^{2/3}}e_j \ln' \mathrm{Ai}\left(\frac{2^{4/3}}{y^{2/3}}\zeta\right)\right]\right\}d\zeta \qquad (12.14)$$

where $\frac{1}{2}\varepsilon$ is the deviation from the critical scattering angle [cf. (12.6)],

$$x \equiv 2\sqrt{2\pi}\,\varepsilon/\eta, \quad y \equiv (4/M')^{3/4}\sqrt{N}\beta^{-1/4} \qquad (12.15)$$

If $\beta$ is so large that $y \ll 1$, the argument of the Airy function in (12.14) is large over most of the range of integration, and one can employ the asymptotic approximation (10.15), so that (12.14) becomes

$$S_{j,\mathrm{TR}}^{(-)} \approx \frac{e^{-i\pi/4}}{2\sqrt{\pi}} \beta\, e^{-2i\beta(M'-N\varepsilon)} \int_0^\infty \exp\left[-i\left(\zeta^2 - x\zeta + e_j y\sqrt{\zeta}\right)\right]d\zeta \quad (12.16)$$

The last (square-root) term in the exponent is the phase associated with the reflection amplitude $R^{(j)}$. In the approximation (12.16), this amplitude is given by (12.5), i.e., it is the Fresnel total reflection amplitude at a *plane* interface, identified with the tangent plane at the point of incidence. Thus, in (12.16), the only effect of the surface curvature that is taken into account is the trivial one of producing a spread in the angles of incidence [Sec. 8.7(v)]: the dynamical effects due to the curvature (centrifugal barrier) on reflection are neglected. For this reason, we call the approximation (12.16) the *planar reflection limit*.

Pearcey's integral $P(x,y)$ is defined by (Pearcey 1947)

$$P(x,y) \equiv \wp(x,y) + \wp(x,-y), \qquad \wp(x,y) \equiv \int_0^\infty \exp\left[i\left(t^4 + xt^2 + yt\right)\right]dt \quad (12.17)$$

in terms of which, after the change of variable $\zeta = t^2$, (12.16) becomes

$$S_{j,\text{TR}}^{(-)} \approx \frac{e^{-i\pi/4}}{2\sqrt{\pi}} \beta e^{-2i\beta(M'-N\varepsilon)} \frac{\partial \wp^*}{\partial y}\left(-x, e_j y\right) \quad (12.18)$$

where the star denotes the complex conjugate. Thus, the planar limit of the Pearcey–Fock integral (12.13) is related with Pearcey's integral, which is the diffraction integral associated with the *cusp catastrophe*, a caustic resulting from confluences of up to three stationary-phase points (Berry & Upstill 1980).

### The Goos–Hänchen shift

In the present situation, only one stationary-phase point is relevant. If one (erroneously) treats the square-root term in the integrand exponent in (12.16) as 'slowly-varying' for $y \ll 1$, taking its value [cf. (12.5)] at the geometrical-optic stationary-phase point $\zeta = x/2$, the result is a Fresnel integral similar to that found in the physical optics approximation, where a similar assumption is made.

The square-root term cannot be treated as slowly-varying, because its derivative diverges at the lower limit of the integral (12.16): this is a manifestation of the 'weak caustic' singularity at the critical angle in the geometrical-optic and WKB approximations. If one computes the stationary-phase point taking into account the contribution from the square-root term, the geometrical-optic relationship between the direct-reflection scattering angle $\theta$ and the angle of incidence $\theta_1$ is modified to (Fiedler-Ferrari *et al.* 1991)

$$\theta = \pi - 2\theta_1 - \delta\theta_j^{\text{GH}}, \qquad \delta\theta_j^{\text{GH}} \approx -\frac{e_j}{\beta}\sqrt{\frac{2N}{M'^3 \varepsilon}} \quad (12.19)$$

provided that $|\delta\theta_j^{\text{GH}}| \ll \varepsilon$.

As is illustrated in fig. 12.6, $\delta\theta_j^{\text{GH}}$ represents an additional angular displacement of the reflected ray, as compared with geometrical reflection. We call it the *spherical Goos–Hänchen angular shift*, by analogy

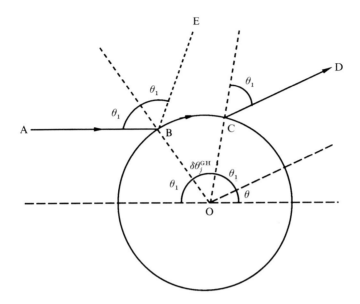

Fig. 12.6.  A ray AB with angle of incidence $\theta_1 > \theta_c$ is not geometrically reflected as BE: it tunnels into the sphere and travels the additional arc BC before reemerging at CD, giving rise to  the spherical Goos–Hänchen angular displacement $\delta\theta_j^{GH}$ (after Fiedler-Ferrari *et al.* 1991).

with the well-known Goos–Hänchen lateral shift in total reflection at a plane interface (Lötsch 1971).

The origin of the Goos–Hänchen shift is the tunneling of light into the optically rarer medium in situation 3, fig. 12.5. The angular displacement can be obtained by taking twice the derivative of the total reflection phase shift with respect to the angular momentum $\lambda$ [cf. (1.26)], a result analogous to Wigner's well-known expression for the time delay in a scattering process (Wigner 1955, Nussenzveig 1972b, Fiedler-Ferrari & Nussenzveig 1981).The expression (12.19) for the angular shift is valid in the planar reflection limit, for shifts much smaller than the deviation from the critical angle.

*Behavior close to the critical angle*

In the immediate neighborhood of the critical angle, the planar-limit results for the full direct-reflection amplitude (partial plus total) can be further simplified, by expanding the Fresnel reflection amplitudes in the integrands into powers of $|\varepsilon|^{1/2} \equiv |\theta - \theta_c|^{1/2}$ [cf. (12.5)]. For the total reflection contribution, this amounts to expanding the exponential of the term with the square root in (12.16) into a power series and integrating term by term. It is found that (Fiedler-Ferrari *et al.* 1991)

$$S_{j,\mathrm{PR}}^{(-)} + S_{j,\mathrm{TR}}^{(-)} \approx -\frac{i\beta}{2}\exp\left(-2i\beta\sin\frac{\theta}{2}\right)$$

$$\times\left[1 - 2^{3/2}\frac{\sqrt{N}}{M'}e_j\left(\frac{M'}{\beta}\right)^{1/4}P\left(-\frac{x}{2}\right) + O\left(\beta^{-1/2}\right)\right] \quad (12.20)$$

where $x$ is defined by (12.15) and

$$P(w) \equiv \frac{e^{i\pi/4}}{\sqrt{\pi}}\left\{\int_{-\infty}^{0}\exp\left[-i(v+w)^2\right]\sqrt{|v|}\,dv\right.$$

$$\left. +i\int_{0}^{\infty}\exp\left[-i(v+w)^2\right]\sqrt{v}\,dv\right\} \quad (12.21)$$

The function $P(w)$ can be expressed in terms of Weber parabolic cylinder functions (Fiedler-Ferrari *et al.* 1991). Its asymptotic behavior for $|w| \gg 1$ is given by

$$P(w) \approx \sqrt{w}\left[1 + \frac{i}{16w^2} + O\left(w^{-4}\right)\right], \qquad\qquad w \gg 1$$

$$\approx i\sqrt{|w|}\left[1 - \frac{\exp(-iw^2)}{2^{3/2}w^2} + \frac{i}{16w^2} + O\left(w^{-4}\right)\right], \quad w \ll -1 \quad (12.22)$$

The dominant term can be obtained from (12.21) by the stationary phase method.

The function $|P(w)|$ is plotted in fig. 12.7, together with its asymptotic limit $\sqrt{|w|}$, which is approached monotonically and rapidly for $w > 0$, whereas the approach is oscillatory and slower for $w < 0$. These features, together with (12.20), already account for the qualitative appearance of the near-critical diffraction pattern (fig. 12.4). Adding the contribution from direct transmission as well, reasonable quantitative agreement is also found for large $\beta$ and small deviations from the critical angle. Note that the scale of the near-critical region appears through the argument $(-x/2)$ of the function $P$ [cf. (12.15)]. For improved approximations, however, one should go back to (12.9) and (12.13).

Also shown in fig. 12.7 is a Fresnel diffraction pattern of the same type as those found in physical optics approximations, showing a substantially different behavior. Indeed, it is found that subcritical and

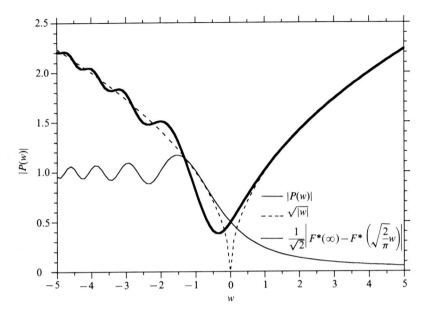

Fig. 12.7. The function $|P(w)|$ and its asymptotic limit $\sqrt{|w|}$, compared with a Fresnel diffraction pattern (after Fiedler-Ferrari *et al.* 1991).

supracritical Fresnel contributions to the amplitudes in the planar limit cancel each other exactly, leaving as the main diffraction correction the term containing the function $P$ in (12.20) (see Sec. 12.7).

The Fresnel pattern may be thought of as the 'Fresnel transform' of the Heaviside step function (geometrical shadow boundary discontinuity), whereas the function $P$, according to (12.21), is the 'Fresnel transform' of the square root function, with its branches defined by

$$\sqrt{w} \to i\sqrt{|w|} \qquad (w < 0) \qquad (12.23)$$

which corresponds to the substitution (12.3) in the Fresnel reflection amplitudes as one goes over from partial to total reflection.

## 12.7  Numerical comparisons

For a quantitative comparison with the CAM approximation, we take numerical Mie results for $N = 0.75$, with farside contributions subtracted out to order $P = 100$ (cf. fig. 12.3). We plot also the WKB approximation and the physical optics approximation.

For $\beta$ = 10 000, the results are shown in fig. 12.8, where the gain factors $G_j(\beta,\theta)$ for both polarizations as well as the cosine of the phase difference $\delta$ defined in (5.3) are plotted.

The geometrical-optic break in slope is still present in the physical optics approximation for $\cos\delta$. In the domain of partial reflection above $\theta_t$, this approximation, in which subcritical reflection is neglected, has much larger errors than the WKB result. Below $\theta_t$, it improves upon the WKB approximation (i.e., it accounts for diffraction) only in the region between the top of the main peak and the second one for the gain factors, but still with considerable errors.

The CAM approximation is in very good agreement with the Mie results: over most of the near-critical region, the deviations are of the same order as the residual fluctuations in the Mie curves, that are not eliminated by the farside subtractions. At the high end of the angular range shown, small residual oscillations are present in the CAM results; slight deviations remain also at the low end of the angular range.

Thus, at this high value of $\beta$, the zero-order transitional CAM approximation already accounts very well for the diffraction effects in near-critical scattering. At $\beta$ = 1000, the agreement is still good, though not quite as good (Fiedler-Ferrari *et al.* 1991); to improve the results, one should include higher-order corrections and make use of uniform asymptotic approximations.

It is instructive to plot separately the magnitudes of the various terms that contribute to the CAM and physical optics approximations. For $N$ = 0.75 and $\beta$ = 10 000, this is illustrated in fig. 12.9, where $g_2 \equiv \sqrt{G_2}$. The curve R is the total reflection contribution to the physical optics approximation, which is a Fresnel pattern, and the curve T is the corresponding contribution from direct transmission, which is just the WKB approximation to this term.

The CAM total reflection contribution TR is displaced from the curve R as a consequence of the Goos–Hänchen shift. The contribution PR from partial reflection, which is of magnitude comparable to TR at $\theta_t$ and becomes dominant above $\theta_t$, is neglected in the physical optics approximation. The Fock-type contribution DT from direct transmission, though small, differs appreciably from the WKB result; in particular, it has a tail extending above $\theta_t$. Its growth as $\theta$ decreases below $\theta_t$ is responsible for the growing amplitude of oscillation apparent in fig. 12.4.

Interference between TR and PR cancels out the relatively large Fresnel-like oscillations about the mean on both sides of $\theta_t$, leaving only considerably smaller 'parabolic-cylinder-like' oscillations for $\theta < \theta_t$ and

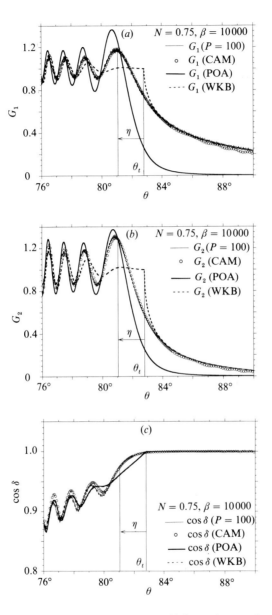

Fig. 12.8. Comparison between subtracted Mie results to order $P = 100$ for N = 0.75, $\beta = 10\,000$ (thin full line), the CAM approximation (open circles), the physical optics approximation POA (thick full line) and the WKB approximation (dashed line): *(a)* Gain factor, polarization 1; *(b)* Gain factor, polarization 2; *(c)* $\cos\delta$ (after Fiedler-Ferrari *et al.* 1991).

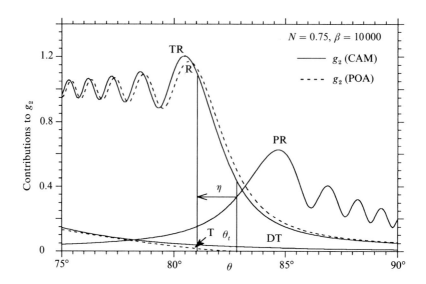

Fig. 12.9. Contributions to $g_2(\beta, \theta) \equiv \sqrt{G_2(\beta, \theta)}$ for $N = 0.75$, $\beta = 10\,000$, for the CAM approximation: from partial reflection (PR), direct transmission (DT) and total reflection (TR); and for the physical optics approximation (POA): from reflection (R) and transmission (T) (after Fiedler-Ferrari *et al.* 1991).

removing the oscillations in the near-critical region for $\theta > \theta_t$. It may be seen in fig. 12.9 that the envelope of the oscillations about the mean in each of the curves TR and PR has an amplitude that is of the same order as the tail of the other curve, leading to this cancellation.

The new diffraction effects in near-critical scattering arise basically from anomalous reflection within this domain. In the planar reflection limit, they are described (as corrections to the WKB results) by the parabolic cylinder function $P(w)$, which is the Fresnel transform of the square root function (12.23). In this limit, it is the singularity associated with the branch point of this function in the Fresnel reflection amplitudes that gives rise to the weak caustic in the geometrical-optic pattern.

However, one must take into account the dynamical effects of the surface curvature, that modify the reflection amplitudes near the critical angle, producing Fock-type effects. Taking them into account in a transitional approximation, we are led to the Fresnel–Fock and Pearcey–Fock integrals, the new diffraction integrals associated with near-critical scattering.

In terms of the short-wave asymptotic behavior of Huygens–Kirchhoff-type integral representations like (10.17), what breaks down in the saddle-point method, in the present case, is not related with the integrand phase $\phi$, as happens in diffraction catastrophes (in spite of the relationship with the cusp catastrophe).

The new features have to do with the integrand amplitude $A$, which, in the saddle-point method, is assumed to be a single-valued holomorphic function within the range of integration. This assumption does not hold in near-critical scattering. In the planar limit, $A$ goes through the square-root branch cut (12.23) within the range of the geometrical-optic saddle point. In the full CAM representation, there is no branch cut, but $A$ is *piecewise analytic*, i.e., it is represented by different analytic functions in different parts of the range: the function $R_{22}^{(j)}$ of (12.12) in the partial-reflection range and the function $R^{(j)}$ in the total reflection range.

As was found in the other semiclassical diffraction effects treated so far, tunneling, manifested through the spherical Goos–Hänchen angular shift, also plays an important role in near-critical scattering.

# 13

# Average cross sections

*Such averaging is required to wash out the sharp bumps.*
(Friedman & Weisskopf 1955)

When absorption within the scatterer is taken into account, the incident beam is extinguished both by scattering and by absorption, and one associates a specific cross section with each of these processes. In applications to radiative transfer, it is important to consider the balance not only of energy but also of momentum: this is related with the radiation pressure cross section.

All of these cross sections, for Mie scattering, are plagued by ripple fluctuations similar to those discussed in Sec. 5.2, complicating considerably their numerical evaluation. These fluctuations are irrelevant for most applications, because they are quenched by size dispersion, and one usually deals with a polydisperse population of scatterers.

Thus, one would like to obtain *average cross sections*, where the average is taken over a size interval just large enough to eliminate the ripple fluctuations, while still allowing slower-scale variations to be resolved. A similar problem occurs in the application of the nuclear optical model to the theory of nuclear reactions (see the above quotation), and one might describe the procedure employed as an 'optical application of the optical model'.

In special circumstances, one may observe, instead of the irregular ripple fluctuations, smooth oscillations of the cross sections about their averages, arising from forward optical glory effects, predicted and explained by complex angular momentum theory.

## 13.1 Efficiency factors

For an absorptive sphere, with $N = n + i\kappa$ ($\kappa > 0$), the scattering cross section, given by (5.4), differs from the *extinction* (total) cross section, given by the optical theorem (9.23), by the *absorption* cross section $\sigma_{abs}$:

$$\sigma \equiv \sigma_{\text{ext}} = \sigma_{\text{sca}} + \sigma_{\text{abs}} \tag{13.1}$$

This expresses the energy balance.

It is customary to employ, instead of the cross sections, the corresponding dimensionless *efficiency factors* (van de Hulst 1957), obtained by dividing the cross sections by the projected area $\pi a^2$ of the sphere:

$$Q_{\text{a}} \equiv \sigma_{\text{a}} \big/ \left( \pi a^2 \right) \tag{13.2}$$

where the index a stands for 'ext', 'sca' or 'abs'. Thus,

$$Q_{\text{ext}} = Q_{\text{sca}} + Q_{\text{abs}} \tag{13.3}$$

For an incident plane wave of intensity $I_0$, the time-averaged momentum in the direction of incidence removed by extinction per unit time is (Jackson 1975)

$$I_0 \, \sigma_{\text{ext}} / c$$

However, through scattering, the corresponding amount restored to the incident beam is

$$\frac{I_0}{c} \int \cos \theta \, \frac{d\sigma_{\text{sca}}}{d\Omega} \, d\Omega \equiv \frac{I_0}{c} \, \overline{\cos \theta} \, \sigma_{\text{sca}}$$

where, in the notation of (5.4),

$$\overline{\cos \theta} \equiv g = \frac{\int_0^{\pi} \left( i_1 + i_2 \right) \cos \theta \sin \theta \, d\theta}{\int_0^{\pi} \left( i_1 + i_2 \right) \sin \theta \, d\theta} \tag{13.4}$$

is called the *asymmetry factor*.

The time-averaged force exerted by the incident radiation on the scatterer (net momentum transmitted to it per unit time) is therefore

$$\frac{I_0}{c} \left( \sigma_{\text{ext}} - \overline{\cos \theta} \, \sigma_{\text{sca}} \right)$$

in the direction of incidence. The corresponding radiation pressure $P$ exerted by the incident beam on the sphere is obtained by dividing by $\pi a^2$, so that it is given by

$$P = \frac{I_0}{c} Q_{\text{pr}} \qquad (13.5)$$

where

$$Q_{\text{pr}} = Q_{\text{ext}} - \overline{\cos\theta}\, Q_{\text{sca}} \qquad (13.6)$$

is called the *radiation pressure efficiency* (van de Hulst 1957). For isotropic scattering, by (13.4), $g = 0$ and $Q_{\text{pr}} = Q_{\text{ext}}$; in the other limiting case of pure forward diffraction, where the angular distribution is concentrated in the forward direction, $g = 1$ and $Q_{\text{pr}} = Q_{\text{abs}}$.

The partial-wave expansion of the absorption efficiency is [cf. (5.6)]

$$Q_{\text{abs}} = \beta^{-2} \sum_{l=1}^{\infty} (l + \tfrac{1}{2})\left\{\left[1 - \left|S_l^{(1)}\right|^2\right] + \left[1 - \left|S_l^{(2)}\right|^2\right]\right\} \qquad (13.7)$$

which vanishes, as it should, for a perfectly transparent sphere ($|S_l^{(j)}| = 1$).

The corresponding expansion for the radiation pressure efficiency may be obtained from (13.6) and (Debye 1909, Irvine 1965)

$$\overline{\cos\theta}\, Q_{\text{sca}} = \beta^{-2} \,\mathrm{Re}\left\{\sum_{l=1}^{\infty} \frac{l(l+2)}{(l+1)}\left[\left(1 - S_l^{(1)}\right)\left(1 - S_{l+1}^{*(1)}\right)\right.\right.$$
$$\left.\left. + \left(1 - S_l^{(2)}\right)\left(1 - S_{l+1}^{*(2)}\right)\right] + \sum_{l=1}^{\infty} \frac{(2l+1)}{l(l+1)}\left[\left(1 - S_l^{(1)}\right)\left(1 - S_l^{*(2)}\right)\right]\right\} \qquad (13.8)$$

where the star denotes complex conjugation.

Plots of the efficiency factors as functions of the size parameter, based on numerical summation of these partial-wave expansions (Irvine 1965) show that all of them undergo rapid ripple fluctuations, similar to those illustrated in fig. 5.2 (though with smaller relative amplitude). A typical example for the extinction efficiency is shown in fig. 13.1 below.

We want to evaluate average efficiency factors

$$\langle Q_{\text{ext}}\rangle, \ \langle Q_{\text{abs}}\rangle, \ \langle Q_{\text{pr}}\rangle \qquad (13.9)$$

where the angle brackets denote a mean over the size parameter, taken over an interval centered at a value $\beta$ and of width $\Delta\beta$, with $\Delta\beta = O(1)$, so that one can wash out the ripple without losing track of variations on a slower scale. A convenient choice is $\Delta\beta \sim \pi$.

## 13.2 CAM theory of average efficiency factors

For $Q_{ext}$, geometrical optics and classical diffraction theory yield as dominant terms those given in (9.24), arising from the direct reflection and direct transmission Debye terms.

The CAM result for these two Debye terms follows from the optical theorem (9.23). In the Fock approximation (cf. Sec. 8.5), it is given by (Nussenzveig & Wiscombe 1980a, Nussenzveig 1990)

$$Q_{ext} = Q_{ext,ax} + Q_{ext,dr} + Q_{ext,ripple} \tag{13.10}$$

where

$$Q_{ext,ax} = -\frac{8}{\beta} \operatorname{Im}\left\{ \frac{N^2}{(N-1)(N+1)^2} \exp[2i(N-1)\beta] \right.$$
$$\times \left[ 1 + \frac{i}{2\beta}\left( \frac{1}{N-1} - \frac{N-1}{N} \right) \right.$$
$$\left. \left. - \sum_{j\geq 1} \frac{(N-1)^{2j+1}}{(N+1)^{2j}} \frac{\exp(4i j\beta)}{[2j-(N-1)]} + O(\beta^{-2}) \right] \right\} \tag{13.11}$$

and

$$Q_{ext,dr} = 2 + 2\operatorname{Re} M_0 \, \gamma^2 + \operatorname{Im}\left( \frac{N^2+1}{M} \right) \gamma^3$$
$$+ \frac{16}{15}\operatorname{Re} M_1 \, \gamma^4 - \frac{1}{6}\operatorname{Im}\left[ M_0 \frac{(N^2+1)}{M^3}(2N^4 - 6N^2 + 3) \right] \gamma^5$$
$$+ \left[ \left( \frac{8}{175} M_2 - \frac{397}{480} \right) + \frac{1}{4}\operatorname{Re}\left( \frac{N^6 - N^2 - 1}{M^2} \right) \right] \gamma^6 + O(\gamma^7) \tag{13.12}$$

with $\gamma = (2/\beta)^{1/3}$, $M = (N^2 - 1)^{1/2}$ and the Fock coefficients $M_0$, $M_1$ and $M_2$ given by (8.30). The $O(\gamma^6)$ term was not included in Nussenzveig & Wiscombe 1980a, where there are also some misprints.

The axial-ray contribution (13.11) includes correction terms to direct transmission and multiply-reflected higher-order terms omitted in (9.24). In the contribution from the direct-reflection Debye term (13.12), the first term arises in equal parts from geometrical reflection and forward diffraction (blocking). The remaining terms arise from edge diffraction (tunneling), in the Fock approximation.

It will be seen in the next chapter that, for $N = 1.33$, one should also include a (small) glory contribution from the $p = 4$ Debye term.

The last term in (13.10), which represents the ripple fluctuations, will also be treated in the next chapter. Since the purpose of averaging is to wash out the ripple, it suffices in this case to take just the sum of (13.11) and (13.12). The $p = 4$ glory term averages out to zero.

An expression for the absorption efficiency according to geometrical optics was obtained by van de Hulst (1946, 1957), by following the fate of the energy flux in an incident pencil of rays, contained within a small solid angle, through its multiple internal reflections. The result is, for incidence of natural light,

$$(Q_{abs})_{go} = \int_0^{\pi/2} \left(1 - e^{-A}\right) \left( \frac{1 - |r_1|^2}{1 - |r_1|^2 e^{-A}} + \frac{1 - |r_2|^2}{1 - |r_2|^2 e^{-A}} \right) \sin\theta_1 \cos\theta_1 \, d\theta_1 \quad (13.13)$$

where absorption is taken into account through the intensity attenuation factor associated with each shortcut through the sphere [cf. fig. 9.5(a)]

$$e^{-A} \equiv \exp\left(-4\kappa\beta\cos\theta_2\right), \quad \kappa = \text{Im } N \quad (13.14)$$

and $\theta_2$ is the angle of refraction corresponding to an angle of incidence $\theta_1$, according to Snell's law applied to the real part of the complex refractive index, $\sin\theta_2 = \sin\theta_1/n$, $n = \text{Re}N$. The quantities $r_1$ and $r_2$ are the Fresnel reflectivities for polarizations 1 and 2, respectively, for the angle of incidence $\theta_1$.

A similar expression for the radiation pressure efficiency was derived by Debye (1909); as shown by van de Hulst (1946, 1957), it has an analogous physical interpretation in terms of geometrical optics.

In a CAM treatment, one can apply the Poisson sum formula to the partial-wave expansions (13.7) and (13.8), but the resulting expressions cannot be analytically extended to the complex $\lambda$ plane because complex conjugation is present.

What is done instead is to begin by averaging the results over $\beta$; in so doing, only the $m = 0$ term survives in the Poisson summation (6.1), and, for (13.7), one is led to replace

$$|S^{(j)}(\lambda,\beta)|^2 \rightarrow \langle |S^{(j)}(\lambda,\beta)|^2 \rangle \quad (13.15)$$

where $S^{(j)}(\lambda,\beta)$ is the $S$-function (9.1) and the angle brackets are defined as in (13.9).

In order to perform the size average, the cylindrical functions in

(9.1) are approximated by suitable asymptotic expansions. In a transitional treatment, these are the Debye asymptotic expansions for the cylindrical functions outside of the edge strip and the Schöbe expansions (Schöbe 1954) within the edge strip. The result is of the form

$$<Q_{abs}> = <Q_{abs}>_F + <Q_{abs}>_{ae} + <Q_{abs}>_{be} \qquad (13.16)$$

In this expression, $<Q_{abs}>_F$ is the contribution from ordinary (Fresnel) reflection and transmission, including internal multireflection. It has exactly the same form as the geometrical-optic result (13.13), but differs from it in two significant respects. The parameter $\theta_2$ is now the *complex* angle of refraction associated with $\theta_1$, given by Snell's law with the *complex* refractive index,

$$\sin\theta_2 = \sin\theta_1/N \qquad (13.17)$$

and the damping exponent $A$ in (13.14) now corresponds to a *complex* shortcut through the sphere:

$$A \rightarrow 4\beta \,\mathrm{Im}\big(N\cos\theta_2 + \theta_2\sin\theta_1\big) \qquad (13.18)$$

The other two terms in (13.16) represent the *edge effects* ('ae' stands for 'above-edge' and 'be' for 'below-edge'). Their functional form is again the same as (13.13), but they now include the dynamical effects of surface curvature on the Fresnel reflectivities, which must be replaced by Fock-type expressions, and the range of angles of incidence now corresponds to the edge strip – including complex angles of incidence above the edge. In the below-edge term, the integrand represents the correction to the Fresnel reflectivities employed in (13.13). Explicit expressions for all terms are given in Nussenzveig & Wiscombe 1980a.

The CAM treatment of the average radiation pressure efficiency is similar. Before applying the Poisson sum formula to (13.8), one expresses all cylindrical functions of index $l+1$ in terms of those of index $l$ and their derivatives, by employing recurrence relations. One also separates out of the $S$-function the contribution from direct reflection. The results (Nussenzveig & Wiscombe 1980a) are analogous to (13.16): again, there is a Fresnel term, representing an improved version of the geometrical-optic result, including the corrections (13.17) and (13.18), and above-edge and below-edge terms, with the same functional form, extended to complex angles of incidence and refraction, including the effects of surface curvature on the reflectivities.

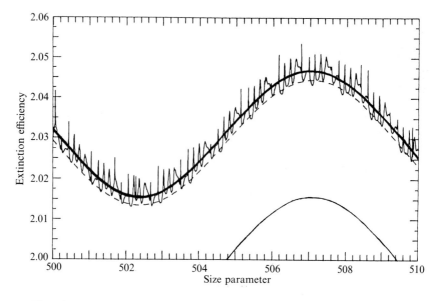

Fig. 13.1. Extinction efficiency for $N = 1.333$ near $\beta = 500$, showing ripple fluctuations. The average behavior is fitted: by CAM (thick-line curve); by classical diffraction and geometrical optics (lower thin-line curve); and by van de Hulst's semi-empirical fit (dashed-line curve) [adapted from Shipley & Weinman 1978].

## 13.3　Numerical results

The extinction efficiency of a perfectly transparent sphere with $N = 1.333$ near $\beta = 500$ is plotted in fig. 13.1, showing ripple fluctuations (Shipley & Weinman 1978). A portion of the curve representing $Q_{ext}$ according to geometrical optics and classical diffraction theory, given by (9.24), appears in the figure (lower thin-line curve): it differs by about 0.03 from $<Q_{ext}>$, corresponding to a relative error of about 1.5% in this range of $\beta$.

　　The dashed-line curve in fig. 13.1 is based on a semi-empirical fit to $Q_{ext}$ due to van de Hulst (1957), which attempts to incorporate the edge effects by adding a term given by $1.84\beta^{-2/3}$ to (9.24). The thick-line curve represents the CAM result, given by the sum of (13.11) and (13.12). It improves the accuracy of the fit to $<Q_{ext}>$, in comparison with geometrical optics plus classical diffraction theory, by more than two orders of magnitude.

　　Detailed comparisons between the numerical Mie results for $<Q_{ext}>$, $<Q_{abs}>$, $<Q_{pr}>$ and the corresponding CAM approximations, over the ranges $10 \leq \beta \leq 5000$, $0 \leq \mathrm{Im}N = \kappa \leq 1$, for $n = \mathrm{Re}N = 1.10$, 1.33, 1.50, 1.90 and 2.50, show greatly improved accuracy in all cases in comparison with previously known expressions [such as (13.13)].

Results for $n = 1.33$ and $10 \leq \beta \leq 1000$ (Nussenzveig & Wiscombe 1980a) may be seen in fig. 13.2. On the left side are shown three-dimensional plots of $<Q_{\mathrm{ext}}>$, $<Q_{\mathrm{abs}}>$, $<Q_{\mathrm{pr}}>$ in this range, with $10^{-5} \leq \kappa \leq 1$. The oscillations in $<Q_{\mathrm{ext}}>$, like those in fig. 13.1, arise from interference between diffracted and transmitted light. They are damped out as $\kappa\beta$ increases: beyond $\kappa\beta \sim 1$, one approaches a flat plateau with the asymptotic limit 2.

For $<Q_{\mathrm{abs}}>$, the rise to values of order unity also takes place beyond $\kappa\beta \sim 1$, but there is a subsequent decrease for large $\kappa\beta$ because the energy gets reflected (metallic limit). As for $<Q_{\mathrm{pr}}>$, its small values for small $\kappa$ and large $\beta$ arise from the dominance of forward diffraction, which returns most of the momentum to the field; again, there is a rise to values of order unity beyond $\kappa\beta \sim 1$, because the effects of reflection and diffraction become comparable.

Level curves for the logarithm of the percentage error of the CAM approximations to the average efficiency factors are shown on the right side of fig. 13.2. Negative values, corresponding to errors $< 1\%$, are shown by dotted lines.

We see from fig. 13.2($b$) that, for $<Q_{\mathrm{ext}}>$, the relative error is already $< 1\%$ at $\beta \gtrsim 15$; it is $< 0.1\%$ at $\beta \gtrsim 70$, $< 0.01\%$ at $\beta \gtrsim 200$ and $< 0.001\%$ at $\beta \gtrsim 1000$. The $O(\gamma^6)$ term in (13.12) was not included in the computations that led to these results; for low values of $\beta$, one should expect that the results would still be considerably improved by employing a uniform CAM approximation.

For $<Q_{\mathrm{abs}}>$ [fig. 13.2($d$)], the relative errors are somewhat larger, and they are the worst for $<Q_{\mathrm{pr}}>$ [fig. 13.2($f$)]: in this case, the error falls below $1\%$ for $\beta \geq 90$. The accuracy increases with $\beta$ and with $n$.

In contrast, previously known approximations, based on geometrical optics and classical diffraction theory, have an accuracy that is almost independent of $n$ and their relative errors only reach $1\%$ at $\beta = 1000$; the errors still range from $0.2\%$ to $0.5\%$ at $\beta = 5000$.

The computing time for the CAM approximations is size-independent and only about twice that taken for geometrical-optic approximations. Relative to exact Mie computations, the gain in computing speed is roughly a factor of order $\beta$: e.g., for the data shown in fig. 13.2, the gain was by about two orders of magnitude.

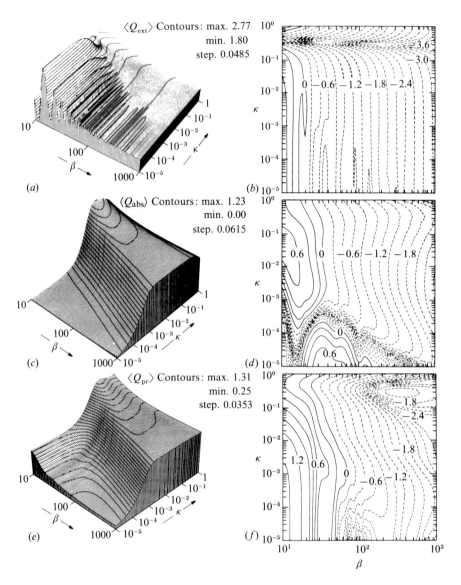

Fig. 13.2. *(a)* Three-dimensional plot of $<Q_{ext}>$ for $n = 1.333$, $10^{-5} \leq \kappa \leq 1$, $10 \leq \beta$ $\leq 1000$. The values of $<Q_{ext}>$ may be read from the contour lines. *(b)* Level curves for the logarithm of the percentage error of the CAM approximation to $<Q_{ext}>$. Negative values (errors < 1%) are shown by dotted lines. *(c)* Same as *(a)* for $<Q_{abs}>$. *(d)* Same as *(b)* for $<Q_{abs}>$. *(e)* Same as *(a)* for $<Q_{pr}>$. *(f)* Same as *(b)* for $<Q_{pr}>$. (from Nussenzveig & Wiscombe 1980a).

The gain in precision relative to previous approximations, besides the improvement in the geometrical-optic type contributions, arises from the edge corrections. The functional form of these contributions, as was mentioned above, is very similar to the geometrical-optic one, extended to complex angles of incidence and refraction. Thus, within the present context, edge diffraction effects may again be interpreted in terms of a kind of analytic continuation of ray optics to complex paths.

## 13.4  Forward optical glory

The meteorological glory treated in Chapter 11 is a backscattering effect. According to the discussion of critical effects given in Chapter 1, glories in the forward direction should also be expected to occur. What are their manifestations in Mie scattering?

As we have seen in Sec. 4.2, the axial focusing amplitude enhancement for a forward glory ray is of order $(\lambda_G)^{1/2}$, where $\lambda_G$ is the glory angular momentum. However, this has to compete with the $O(\beta)$ amplitude enhancement in the forward direction due to diffraction. Thus, to improve its chances of being seen over the dominant forward diffraction peak, a forward glory should satisfy the following requirements (Nussenzveig & Wiscombe 1980b):

(i)  The number of internal reflections of the glory rays should be as small as possible, so as to minimize the internal-reflection damping, as well as the length of the optical path and the associated damping by absorption (for $\kappa = \mathrm{Im}N > 0$).

(ii)  For the same reason, the internal reflectivity should be as large as possible, suggesting near-tangential incidence (within the edge strip). This also maximizes the axial focusing enhancement ($\lambda_G = \beta$).

Under these conditions, the smallest number of internal reflections to produce a real forward glory ray is two, with $n \geq 2$. For $n = 2$, the forward glory corresponds to tangential incidence and the ray path is an inscribed equilateral triangle, the lowest geometric resonance (cf. Sec. 11.4).

If $n$ is less than but close to 2, one has the situation illustrated in fig. 13.3: the triangle does not quite close, there remaining an angular gap

$$\zeta = 6\theta_l - \pi, \qquad \theta_l \equiv \sin^{-1}(1/n) \qquad (13.19)$$

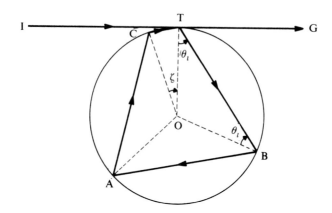

Fig. 13.3. Forward glory path for a tangentially incident ray IT. The portion CT corresponds to a surface wave. Portions of surface-wave paths may also be described around T, B and A, always adding up to the same missing angle $\zeta = 6\theta_l - \pi$, where $\theta_l$ is the critical angle (after Nussenzveig & Wiscombe 1980b).

where $\theta_l$ is the critical angle. The missing angle $\zeta$ is approximately 16° for $n = 1.85$, 10.5° for $n = 1.90$ and 5° for $n = 1.95$.

Similarly to the van de Hulst term in the backward glory, the gap is bridged by surface waves, associated with the $p = 3$ Debye term in the present case. Besides the path shown in fig. 13.3, other possible paths corresponding to the same total missing angle would allow for portions of surface-wave paths described around the points T, B and A.

Interference between the forward glory contribution and other contributions to forward scattering [mainly those contained in (9.24), arising from diffraction and direct transmission] should give rise to a sinusoidal modulation of these contributions in $Q_{ext}$. This indeed appears quite clearly in fig. 13.4, where $N = 1.9 + 10^{-4}$ i.

A small amount of absorption has been added to $n$ in this example in order to damp out other contributions. The dashed-line curve in fig. 13.4 is the CAM asymptotic approximation to $Q_{ext}$ omitting the ripple, given by the sum of (13.11) and (13.12). By comparing fig. 13.4 with fig. 13.1, we see a striking difference: instead of the usual irregular ripple fluctuation around the asymptotic curve, there is a regular, nearly sinusoidal oscillation (apart from minor residual ripple components).

The sinusoidal component represents forward glory undulations, arising from interference between the glory term and the others. The

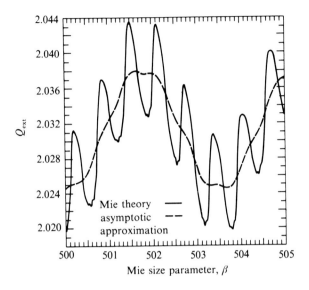

Fig. 13.4. Exact Mie extinction efficiency $Q_{ext}$ (full line) and asymptotic CAM approximation excluding ripple (dashed line) for $N = 1.9 + 10^{-4}i$ and $500 \leq \beta \leq 505$ (after Nussenzveig & Wiscombe 1980b).

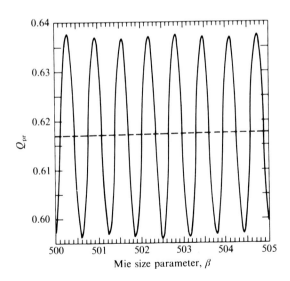

Fig. 13.5. Exact Mie radiation pressure efficiency $Q_{pr}$ (full line) and asymptotic CAM approximation $\langle Q_{pr} \rangle$, averaged over $\Delta\beta = \pi$ (dashed line) for $N = 1.9 + 10^{-4}i$ and $500 \leq \beta \leq 505$ (after Nussenzveig & Wiscombe 1980b).

amplitude of the forward glory contribution is represented by half the peak-to-valley amplitude of the sinusoidal component.

An even more striking manifestation of the forward glory appears in fig. 13.5, which shows the behavior of the Mie radiation pressure efficiency $Q_{pr}$ in the same range (full-line curve), together with the CAM asymptotic approximation to $<Q_{pr}>$ described in Sec. 13.2 (dashed-line curve). The forward-glory oscillations about the average curve are seen to be highly sinusoidal in this case.

A large number of other examples have been plotted, covering the ranges $1.8 \leq n \leq 2.05$, $10^{-7} \leq \kappa \leq 10^{-2}$, $50 \leq \beta \leq 1500$. Forward glory oscillations are seen very clearly in all Mie efficiency factors, including $Q_{abs}$ and $Q_{sca}$. The amplitude of the glory oscillations increases as $\kappa\beta$ decreases, but so also does the admixture with ripple fluctuations, especially for $Q_{abs}$.

From these graphs, one can estimate the period $(\delta\beta)_{Mie}$ of the forward glory oscillations with an accuracy of the order of 0.001. According to CAM theory, the period is given by

$$(\delta\beta)_{CAM} = 2\pi / \left(6\sqrt{n^2 - 1} + \zeta\right) \tag{13.20}$$

which corresponds to the optical path difference associated with the three shortcuts together with the arc CT in fig. 13.3. A comparison between this theoretical period and that measured from the graphs is shown in Table 13.1. The agreement is seen to be excellent.

Table 13.1. *Comparison between Mie and CAM forward glory oscillation periods* (from Nussenzveig & Wiscombe 1980b)

| $n$ | 1.850 | 1.900 | 1.950 | 2.000 |
|---|---|---|---|---|
| $(\delta\beta)_{CAM}$ | 0.653 | 0.636 | 0.620 | 0.605 |
| $(\delta\beta)_{Mie}$ | 0.653 | 0.636 | 0.620 | 0.604 |

Another characteristic feature of the forward glory is that it leads to the appearance of cross-polarization (cf. Sec. 11.2) in near-forward scattering. Indeed, while the forward diffraction peak is unpolarized, the forward glory, as is typical for surface-wave contributions, is dominated by parallel (electric) polarization. This leads to an oscillatory behavior of the degree of linear polarization, that should be particularly noticeable near the first diffraction minimum (Nussenzveig & Wiscombe 1980b).

Similar forward glory effects are found near the next geometrical resonance (inscribed square), that takes place when $n$ goes through $\sqrt{2}$. They are associated with the $p = 4$ Debye term. While the additional internal reflection increases the damping of the forward glory contribution, this refractive index range is more readily accessible.

It will be seen in the next chapter that one can also detect forward glory effects arising from the $p = 4$ Debye term in the extinction efficiency of water droplets. The missing angle $\zeta$ that has to be bridged by surface waves is approximately 30° (twice the van de Hulst gap of fig. 4.4), so that the surface-wave contribution is small. However, real forward glory rays with $p = 4$ also contribute: this is the smallest $p$ for which they exist in this case. The total forward glory contribution, though small, must be added to (13.11) and (13.12) in order to obtain an accurate fit to $Q_{ext}$.

# Orbiting and resonances

*The phenomenon of resonance is, however, most striking
in cases where a very accurate equality of periods is
necessary in order to ellicit the full effect.*
(Lord Rayleigh 1877)

We now deal with the last of the critical effects listed in Chapter 1: orbiting, which, as was mentioned in Sec. 1.3, is often associated with the presence of resonances. This is also true here.

The main feature of Mie scattering that we have not treated so far, the existence of sharp ripple fluctuations for $\mathrm{Re}N > 1$, is due to resonances. These resonances, in the idealized model of a perfectly transparent sphere, grow arbitrarily sharp with increasing size parameter; in real life, they can still be extremely narrow, leading to a variety of interesting applications that will be described in Chapter 15.

The origin and physical interpretation of the resonances are greatly clarified by employing the effective potential picture. The resonances appear as 'quasibound states of light', associated with orbiting-like paths of light rays around the scatterer. Tunneling through the centrifugal barrier plays an essential role in determining the resonance widths and in explaining the 'sensitivity to initial conditions' that characterizes the ripple fluctuations.

CAM theory describes the resonances in terms of families of *Regge trajectories*, a concept that has found important applications in high-energy physics (Collins 1977). It allows one to determine the resonance positions and widths with high accuracy.

The resonance contributions correspond to residues at the Regge poles, following their trajectories. By adding them to the background that results from the other effects already treated by CAM methods, one can finally obtain complete fits of the Mie cross sections, including the ripple.

The broader resonances just below the barrier top, which are least sensitive to parameter deformations, give a substantial contribution to backscattering, which must be added to the background terms discussed in Chapter 11 in order to complete the treatment of the meteorological glory.

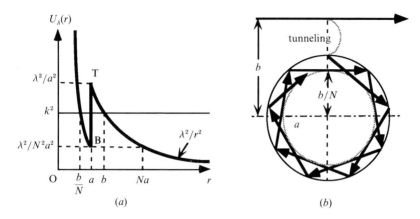

Fig. 14.1. *(a)* Effective potential for a transparent sphere, $N > 1$. The 'energy level' $k^2$ is such that $\beta < \lambda < \alpha$. There are three turning points: $r = b/N$, $r = a$ and $r = b$, where $b = \lambda/k$. *(b)* The corresponding incident ray, with impact parameter $b$, tunnels through the centrifugal barrier to the surface and is multiply reflected inside, travelling beyond the critical angle, at a distance $r = b/N$ from the center.

## 14.1  Effective potential and resonances

The effective potential for a perfectly transparent sphere with $N > 1$, already represented in fig. 9.1, is reproduced in fig. 14.1*(a)*. In the region between points T and B, one has a potential pocket [of depth $(N^2 - 1)k^2$] surrounded by a barrier, a situation which, as is well known (Berry & Mount 1972), can lead to the formation of sharp resonances.

The 'energy level' $k^2$ shown is such that $\beta < \lambda < \alpha$; thus, by the localization principle, it corresponds to an impact parameter $b$ such that $a < b = \lambda/k < Na$, i.e., to an incident ray passing outside of the sphere. It leads to three radial turning points: $r = b/N$, $r = a$ and $r = b$.

As illustrated in fig. 14.1*(b)*, getting across the forbidden region between $a$ and $b$ requires *tunneling through the centrifugal barrier*. Once inside, multiple reflections between $r = b/N$ and $r = a$ take place, corresponding to rays travelling internally beyond the critical angle and being *almost* totally reflected: there is a small leakage to the outside at each internal reflection, again by tunneling through the barrier.

Within narrow neighborhoods of 'resonance energies', slope matching of oscillatory-type solutions in the allowed regions and exponential-type ones in the forbidden regions, starting from regular behavior at the origin, leads to very large ratios of internal to external amplitudes, as is schematically illustrated in fig. 14.2.

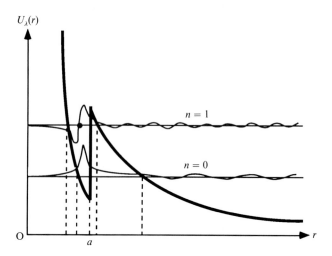

Fig. 14.2.  Schematic representation of resonant wave functions for the resonances $n = 0$
(no nodes within the well) and $n = 1$ (one node ● within the well).

The resonances may be thought of as 'quasibound' states, that would become bound in the limit of zero leakage through the barrier. They are characterized by a 'family number' $n$ (= 0, 1, 2, ...), the number of nodes of the radial wave function within the well in this limit (fig. 14.2). The lower $n$ is, the deeper the quasibound state lies.

Because of the leakage, the resonant states are not stationary: they decay, but a very small barrier transmissivity leads to a very long lifetime, corresponding to many internal reflections. In the ray picture [fig. 14.1(*a*)], this is analogous to *orbiting*. In view of the high angular momenta carried by the resonant states and their confinement to a relatively narrow neighborhood of the surface, they also resemble *whispering-gallery modes* in acoustics (Rayleigh 1877, Walker 1978b).

Because of the dependence of such resonances on the shape of the scatterer, they have become known as 'morphology-dependent resonances' (Barber & Chang 1988). They provide a simple illustration of *shape resonances* (Gustafsson 1983).

For a narrow resonance, the amplitude of internal excitation by an incident plane wave through tunneling [fig. 14.1(*b*)] is very small, but the decay amplitude is also very small, so that the internal field amplitude can build up to very large values as compared with the external one within the large resonance lifetime.

One can also describe this process as an excitation of the 'free' (or 'natural') modes of oscillation associated with the scatterer, defined by the condition that they contain only outgoing radiation. This non-selfadjoint boundary condition yields *complex eigenfrequencies*, that correspond to the poles of the *S*-function. In an energy picture, their real part gives the resonance energies and their imaginary part determines the resonance widths or lifetimes: considerable care is required in the physical interpretation (Nussenzveig 1972a). An early discussion of the free modes of vibration of a dielectric sphere was given by Debye (1909).

## 14.2 The poles of the *S*-function

The poles of the function $S^{(j)}(\lambda,\beta)$ are the roots of the complex transcendental equation [cf. (9.1)–(9.3)]

$$\ln'H_\lambda^{(1)}(\beta) + \frac{1}{2\beta} = Ne_j\left[\ln'J_\lambda(\alpha) + \frac{1}{2\alpha}\right] \qquad (14.1)$$

For real (physical) $\beta$, the poles in the complex $\lambda$ plane are the Regge poles of Sec. 9.2. For physical $\lambda = l + \frac{1}{2}$, one can also solve (14.1) with respect to $\beta$, to find the corresponding poles in the complex $\beta$ plane.

We are interested in the Regge poles located in region 1 of fig. 9.2, with real parts between $\lambda = \beta$ and $\lambda = \alpha$. A first approximation to their location (Nussenzveig 1969a, Guimarães & Nussenzveig 1992) is obtained by employing the Debye asymptotic expansions of the cylindrical functions in (14.1), including a subdominant contribution to the Hankel function in order to get the imaginary part of the poles, which is exponentially small. This also corresponds to a WKB-type approximation (Probert-Jones 1984).

In terms of the effective radial wave number within the well,

$$k_{\mathrm{eff}}(r) = \sqrt{N^2k^2 - \frac{\lambda^2}{r^2}} \qquad (14.2)$$

where $N$ is assumed to be real, the real part of the poles, in the lowest order of approximation, is found to be determined by

$$2\int_{\lambda/Nk}^{a} k_{\mathrm{eff}}(r)\,dr = (2n+1)\pi \quad (n = 0,1,2,...) \qquad (14.3)$$

which is analogous to a Bohr-Sommerfeld quantization condition (Berry & Mount 1972) for 'bound' states of light. The left-hand side represents a radial phase integral between the inner turning point $\lambda/Nk$ and the surface $r = a$. The integer $n$ is the 'family number' of fig. 14.2.

The imaginary part of the poles, in the same approximation, is proportional to the centrifugal barrier penetration factor

$$\exp\left[-2\,\Psi(a,\lambda/k)\right] \equiv \exp\left[-2\int_a^{\lambda/k} \sqrt{\frac{\lambda^2}{r^2} - k^2}\, dr\right] \qquad (14.4)$$

where $\Psi$ represents a radial phase integral between the surface $r = a$ and the outer turning point $\lambda/k$ (fig. 14.2).

The accuracy of the lowest-order WKB-type approximation to the poles can be improved by including higher-order WKB corrections and iterating the transcendental equation for their real part. However, for very narrow resonances, the determination of the real part with an accuracy comparable to the resonance width requires using the uniform asymptotic approximations to the cylindrical functions in (14.1). The results agree with those obtained by more laborious numerical techniques (Conwell, Barber & Rushforth 1984, Hill *et al.* 1985).

For varying $\beta$, each Regge pole $\lambda_{nj}(\beta) = \text{Re}\lambda_{nj}(\beta) + i\,\text{Im}\lambda_{nj}(\beta)$ describes a *Regge trajectory* (de Alfaro & Regge 1965). Two families of Regge trajectories, $n = 0$ and $n = 1$ ($N = 1.33$, both polarizations) for $\beta$ in the range from 50 to 60, are represented in fig. 14.3, where $\text{Re}\lambda_{nj}$ and $\log_{10}\text{Im}\lambda_{nj}$ are plotted against $\beta$ (Guimarães & Nussenzveig 1992). Both behave linearly within this range, so that the trajectories in the $\lambda$ plane tend to approach the real axis exponentially as $\beta$ increases, reflecting the exponential behavior of the barrier transmissivity (14.4).

A resonance with polarization $j$ in the $l$th partial wave arises when $\text{Re}\lambda_{nj}$ intersects the physical value $\lambda = l + \frac{1}{2}$. The resonance width is determined by $\text{Im}\lambda_{nj}$. Thus, each Regge trajectory gives a unified description of all resonances with the same family number $n$. The locations of the $n = 0$ and $n = 1$ resonances (with $l = 65$ and $l = 71$) for both polarizations, in the interval [58.2, 59.0] are marked by open circles in fig. 14.3.

The physical interpretation of the resonance widths is analogous to the well-known interpretation of resonances in potential scattering (Blatt & Weisskopf 1952): the lifetime of a resonant state (inverse of the level width) is of the order of the period of internal motion in the well multiplied by the average number of internal reflections required for transmission to the outside region. This number is the inverse of the combined transmissivity, resulting from the transmissivity (14.4) of the centrifugal

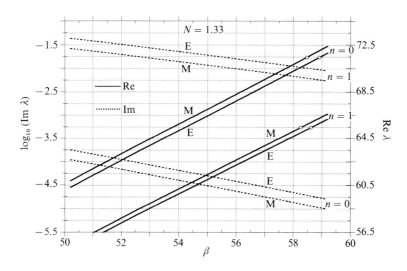

Fig. 14.3. Lowest $n$ Regge trajectories for $N = 1.33$ and $50 < \beta < 60$, for both polarizations E and M. Re $\lambda$: right scale; $\log_{10}$ Im $\lambda$: left scale; o: resonance locations ($l = 65, 71$) for $58.2 \leq \beta \leq 59.0$ (after Guimarães & Nussenzveig 1992).

barrier and that of the potential step at $r = a$.

The dominant effect on the resonance widths is that of tunneling through the centrifugal barrier. As may be seen in fig. 14.2, the height and width of the barrier that must be overcome both increase as $\lambda$ increases or $n$ decreases, explaining the exponentially fast approach of the Regge trajectories to the real axis (that would lead to extremely small resonance widths in the absence of losses), and the relatively large width variation between trajectories with $n$ values differing by unity (fig. 14.3).

An equivalent physical interpretation of the resonances directly in terms of the Regge poles can be given, employing the decomposition (7.22) into running angular waves. The imaginary part of the poles is then related to a 'life-angle' rather than a lifetime (Nussenzveig 1972a).

## 14.3 Resonance and background contributions

How does the resonance contribution to Mie scattering arise in the CAM picture? The treatment of all effects considered so far, for $N > 1$, was based on the Debye expansion (9.5) of the $S$-function. We have seen that, usually, only a small number of terms of this expansion need to be taken into account.

The Regge pole contributions are contained in the *remainder term* $\Delta S_{jp}(\beta,\theta)$ of (9.10), arising from the remainder term $\Delta S_p^{(j)}(\lambda,\beta)$ in (9.5). The poles correspond to $\rho^{(j)}(\lambda,\beta) = 1$ in (9.8), so that they are *singularities of the Debye expansion*, where this expansion breaks down.

Since contributions to the remainder term from $\lambda < \beta$ outside of the edge strip are strongly damped by multiple internal reflections, the only significant contributions arise from $\lambda \gtrsim \beta$, where the path of integration in the Poisson transform can be deformed into the upper half of the $\lambda$ plane, picking up the residues at the Regge poles with $\mathrm{Re}\,\lambda \gtrsim \beta$ (cf. fig. 9.2). These are the poles associated with resonances below the top of the centrifugal barrier, in the range between T and B in fig. 14.1.

The contribution from a resonance associated with the Regge pole $\lambda_{nj}(\beta)$ to the extinction efficiency, obtained through the optical theorem (9.23), is

$$Q_{\mathrm{ext},\lambda_{nj}}(\beta) = -\frac{2\pi}{\beta^2}\,\mathrm{Im}\left\{\frac{\lambda_{nj}(\beta)\exp\left[i\pi\lambda_{nj}(\beta)\right]}{\cos\left[\pi\lambda_{nj}(\beta)\right]}\left[\mathrm{Re}\,s\,S^{(j)}(\lambda,\beta)\right]_{\lambda=\lambda_{nj}}\right\} \quad (14.5)$$

The last square bracket denotes the residue of the $S$-function at the pole.

The resonant behavior originates from the cosine in the denominator, which becomes small at a narrow resonance (when $\lambda_{nj}$ is close to a half-integer), giving rise to a resonance peak. In the immediate neighborhood of the peak, (14.5) yields the well-known Breit–Wigner (Lorentzian) resonance shape (Nussenzveig 1972a).

The peak contribution to $Q_{\mathrm{ext}}$ is given by $2(2l + 1)/\beta^2$, in agreement with the peak (resonant) contribution from a given partial wave and polarization (van de Hulst 1957). Since $l = O(\beta)$, we see that this contribution is $O(\beta^{-1})$, so that it becomes smaller and smaller, relative to the asymptotic limit 2, as $\beta$ increases.

What about the background contribution? This arises from the Debye terms: the contributions from the terms $p = 0$ and $p = 1$, in the Fock approximation, are given by (13.10)–(13.12); an improved result for them may be obtained by employing uniform approximations similar to those discussed in Chapter 8. As will be seen in Sec. 14.4, however, an additional forward glory background term to $Q_{\mathrm{ext}}$ should be included.

Corresponding results are found at nonforward angles. The resonance contributions are similar to (14.5), with residues now including Legendre functions $t_{\lambda_{nj}-\frac{1}{2}}(\pm\cos\theta)$, $p_{\lambda_{nj}-\frac{1}{2}}(\pm\cos\theta)$. For backscattering, this gives rise to an axial focusing enhancement of these contributions by a factor $O(\beta^{1/2})$, consistent with their above-edge origin. The peak contribution to the backscattering gain is of order $\lambda^2/\beta^2 = O(1)$, so that the ratio of resonances to background is strongly enhanced.

The background contribution to the scattering amplitudes, arising from the terms in the Debye expansion, has been discussed in the preceding chapters: it corresponds to the WKB approximation corrected by diffraction effects (penumbra, rainbow, forward or backward glory, ...).

### 14.4   CAM theory of the ripple

The sharp ripple fluctuations in Mie scattering are due to resonances (Kerker 1969, Barber & Chang 1988). In the immediate neighborhood of a peak (within its width), each resonance is associated with a single partial wave. However, resonance contributions in CAM theory, such as (14.5), are associated with *Regge trajectories*, which, as will now be seen, may also strongly affect the behavior of cross sections outside of the resonance peaks.

By plotting the CAM approximation to $Q_{ext}$, with the background given by the Debye terms $p = 0$ and $p = 1$, i.e., by the sum of (13.11) and (13.12) (in Fock approximation), one finds that it differs from the Mie result by a small sinusoidal residual, reminiscent of the forward glory oscillations discussed in Sec. 13.4. Indeed, the main contribution to this difference (Guimarães & Nussenzveig 1992) is due to a $p = 4$ forward glory term.

As illustrated in fig. 14.4, the $p = 4$ forward glory for $N = 1.33$ contains both a real glory-ray component, with an angle of incidence $\theta_1 \approx 49.5°$ [fig. 14.4(*a*)], and a surface-wave component, similar to that of fig. 13.3, where $\zeta \approx 30°$ is the total angular gap, the sum of arcs that may be described at each vertex [fig. 14.4(*b*)]. This surface-wave contribution is the forward counterpart of the van de Hulst backward glory term of fig. 4.4, with twice as large an angular gap.

A CAM fit to $Q_{ext}$, including both background and Regge terms (Guimarães & Nussenzveig 1992), with $58.2 \leq \beta \leq 60.0$ (an interval of about one quasiperiod), for $N = 1.33$, is shown in fig. 14.5. Within this range, the only resonances below the barrier top arise from the Regge trajectory families $n = 0, 1, 2$, one of each for each polarization. Their positions are indicated in the figure (for $n = 0, 1$, cf. also fig. 14.3). The pole contributions (14.5) are evaluated by employing the uniform asymptotic approximations to the cylindrical functions: this is essential to obtain accurate results.

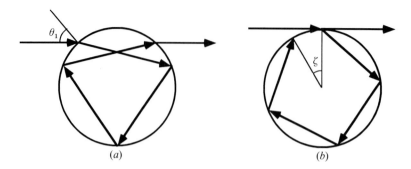

Fig. 14.4. Forward glory paths for the $p = 4$ Debye term, for $N = 1.33$, from: *(a)* real glory rays, with $\theta_1 \approx 49.5°$; *(b)* surface waves, with missing angle $\zeta \approx 30°$ (the sum of arcs that may be described at each vertex) (after Guimarães & Nussenzveig 1992).

The background contribution is plotted in fig. 14.5 both omitting and including the $p = 4$ term: the latter is employed in the CAM fit. The oscillatory character of the forward glory term is clearly seen; the contributions from real glory rays and from surface waves have comparable amplitudes here.

The contribution from the $j = 1$, $n = 2$ Regge trajectory, which gives rise to the $M^2_{61}$ broad resonance, is separately plotted (offset by 2). It is remarkable that it turns negative between consecutive physical points, producing large 'antiresonant' drops of $Q_{ext}$ below the background. This shows that the Regge trajectory contributions (14.5) behave quite differently from that of a single (resonant) partial wave to $Q_{ext}$, which is always positive, as well as from the equally positive Breit–Wigner approximation, which they reproduce in the immediate neighborhood of a narrow resonance peak. The contributions from the broadest resonances, $M^2_{61}$ and $E^2_{61}$, are merged together. Cancellation effects among positive and negative Regge contributions, arising from the phase factor in (14.5), explain why the average $<Q_{ext}>$ is so well approximated by the background (cf. fig. 13.1 and Sec. 13.3).

We see that the fit is very good over the whole range, with a typical accuracy of 0.05% to 0.075%, only becoming worse (by a factor of about 2) near the broad resonances. The accuracy, which is consistent with that of the asymptotic approximations (to order $\beta^{-2}$) employed, is considerably improved by including the $p = 4$ term. The deviations may be attributed partly to the use of Fock rather than uniform approximations for the background and (mostly) to the neglect of corrections of order $(\text{Im } \lambda_{nj})^2$ in evaluating contributions from the broad resonances.

Several features of the ripple pattern (Probert-Jones 1984), derived

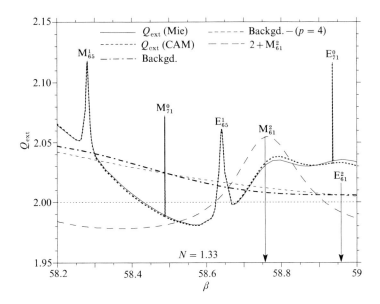

Fig. 14.5. Extinction efficiency for $N = 1.33$: comparison between Mie results and a CAM approximation. Resonance positions are indicated. The background contribution, both excluding and including $p = 4$, is shown. The contribution, offset by 2, from the ($j = 1$, $n = 2$) Regge trajectory ($2 + M^2_{61}$), is also plotted: portions below 2 represent negative contributions (after Guimarães & Nussenzveig 1992).

from the WKB-type approximation to the poles mentioned in Sec. 14.2, may be verified in fig. 14.4. A rough estimate of the spacing $\delta\beta$ between successive passages of a Regge trajectory near physical values of the angular momentum (recurrence of a resonance in consecutive partial waves) is given by $\delta\beta \equiv \beta'_{nj}(l + 1) - \beta'_{nj}(l) \sim \tan^{-1}M/M$, $M = (N^2 - 1)^{1/2}$ (Chylek 1976, 1990). For $N = 1.33$, one gets $\delta\beta \approx 0.821$, a value very close to the quasiperiod of the glory background found in (11.29), leading to a quasiperiodic structure of the full pattern.

One also finds that the resonance width increases very rapidly with the family number $n$, reflecting the rapid variation of the barrier transmissivity with the level height in fig. 14.2. Typically, within the range shown in fig. 14.5, the ratio of widths between resonances ($n + 1$) and $n$ is of order 10 to $10^2$ (cf. also fig. 14.3).

Polarization 1 (M) resonances always appear before polarization 2 (E) resonances with the same angular momentum and family number, and they are narrower. The total density of resonances located below the

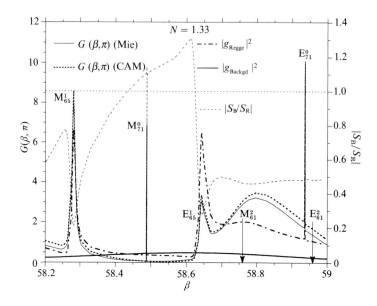

Fig. 14.6. Backward gain for $N = 1.33$: comparison between Mie results and a CAM approximation including $p = 0, 2, 11$ in the background. The corresponding results $|g_{Regge}|^2$ for the Regge term alone and $|g_{Backgd}|^2$ for the background term alone are also plotted (left scale). The ratio $|S_B/S_R|$ of background to Regge backward scattering amplitudes is plotted (right scale) (after Guimarães & Nussenzveig 1992).

top of the centrifugal barrier (per unit size parameter interval) grows approximately linearly with $\beta$ (Hill & Benner 1986).

The most striking manifestation of the ripple fluctuations, as illustrated by fig. 5.2, occurs in backscattering, presenting the greatest challenge for a theoretical explanation and fit. The CAM treatment of the background corresponds in this case to the theory of the glory developed in Chapter 11, and the resonance contribution from Regge trajectories is similar to (14.5).

A CAM fit to the backward gain $G(\beta,\pi)$ within the same range as in fig. 14.5 is shown in fig. 14.6. The background contribution is approximated by the sum of the direct reflection Debye term $p = 0$ with the two leading Debye contributions to the glory in this range, $p = 2$ and $p = 11$. Neglect of other Debye terms represented in fig. 11.7 limits the accuracy of this approximation.

The ripple fluctuations in $G(\beta,\pi)$ are much larger than those in $Q_{ext}$: the gain varies by over two orders of magnitude across this range. As we have seen in Sec. 11.3, the geometrical-optic (WKB) contribution to

$G(\beta,\pi)$ is negligible: the backward gain arises almost entirely from complex paths. Even though several significant Debye contributions to the background, as well as corrections to the broad resonances, are not included, the quality of the CAM fit in fig. 14.6 is still fairly good, with typical errors of about 20% (increasing to twice as much in ranges where $G$ is small). In view of the omissions, this is a very reasonable result.

Although the resonances appear to be overwhelmingly dominant in backscattering (Probert-Jones 1984), this is not so: as shown in fig. 14.6, the resonant (Regge) contribution $|g_{\mathrm{Regge}}|^2$ alone gives a much worse fit, with errors reaching several hundred percent, even at resonance peaks. The same applies to a plot of the background contribution $|g_{\mathrm{Backgd}}|^2$ by itself. The net gain, $G \equiv |g_{\mathrm{Regge}} + g_{\mathrm{Backgd}}|^2$, contains strong interference effects, destructive as well as constructive, between the two terms. The ratio $|S_{\mathrm{B}}/S_{\mathrm{R}}|$ of background to Regge backward scattering amplitudes, also plotted in fig. 14.6, is $\gtrsim 0.5$ over most of the interval.

The 'sensitivity to initial conditions' that is characteristic of the ripple fluctuations (contributing to the quasichaotic features seen in fig. 11.2) reflects the sensitivity of the resonance features to the shape of the effective potential. The resonance widths, determined by the tunneling transmissivity, have an exponential sensitivity to changes in barrier height and breadth: the exponent in (14.4) plays a role analogous to that of the Lyapunov exponent (Bergé *et al.* 1987).

In particular, for large $\beta$, low-lying resonances, too narrow to be resolved, e.g., in fig. 5.2, tend to be washed out by small perturbations (residual absorption, size and shape fluctuations, ...). The most robust resonances are the broadest ones, located near the top T of the barrier in fig. 14.1, that are connected with above-edge incidence within or close to the edge strip. These are the resonances mainly responsible for the ripple usually resolved in numerical Mie calculations. They also contribute substantially to the glory. In the treatment of the glory given in Sec. 11.5, it was possible to reproduce a ripple spike, such as that shown in fig. 11.7(c), only by summing over an infinite subset of terms in the Debye series (associated with the orbiting period). Such an infinite summation approximates a Regge trajectory contribution.

The CAM treatment of Mie resonances remains valid in the presence of material losses (Im $N > 0$). It also yields accurate approximations to the behavior of the resonant field within the sphere, which, for very narrow resonances, builds up to very large values that may produce nonlinear optical effects (cf. Sec. 5.3); observations and applications are discussed in Sec. 15.5.

# 15

# Macroscopic applications

*Nature uses only the longest threads to weave
her patterns, so each small piece of her fabric
reveals the organization of the entire tapestry.*
(Feynman, Leighton & Sands 1964)

In this and in the following chapter, we present a very sketchy and incomplete survey of selected applications and illustrations of semiclassical diffraction effects in various branches of physics. The purpose is to furnish a sample of the variety of areas in which manifestations of similar phenomena are present, emphasizing recent developments, as well as to provide some guidance to the literature.

## 15.1  Recent applications of Mie scattering

The reader is referred to previous monographs (van de Hulst 1957, Kerker 1969, McCartney 1976, Bohren & Huffman 1983) for surveys of traditional applications of light scattering by spherical particles. Detailed discussions of coronae, rainbows and glories in meteorological optics are given in Linke & Möller 1961. We review a couple of more recent applications.

*Glare spots*

When a suspended liquid droplet is illuminated by a parallel light beam and viewed from a relatively close distance (such that it subtends an angle of a few degrees at the observation point), a number of glare spots are seen on its surface; as the observer moves around the droplet, glare spots merge and become colored when rainbow angles are crossed (Tricker 1970, Walker 1976, 1977, 1978). Glare spots are clearly visible in the color pictures of droplets reproduced in Qian *et al* (1986).

The observer's distance, though close, is still far enough that the field is given by the far-zone expressions (5.1). Let $r$ be the distance between the droplet and the lens (observer's eye or camera), focused on

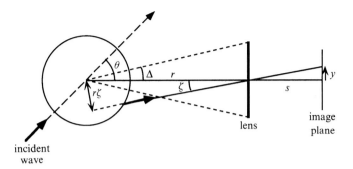

Fig. 15.1. Geometry for observation of glare spots. $\theta$ = scattering angle; $\Delta$ = half-opening angle of lens; $\zeta \approx y/s$ = direction associated with image point $y$.

the droplet, and $\Delta$ the half-angle subtended by the lens, as seen from the droplet (fig. 15.1).

The lens receives contributions from a range of angles between $\theta + \Delta$ and $\theta - \Delta$, centered around the scattering angle $\theta$. We denote by $\zeta$ the direction, as seen from the lens center, associated with an image point with ordinate $y$; we have $\zeta \approx y/s$, where $s$ is the image distance (fig. 15.1).

The field amplitude at $y$ (omitting spherical wave propagation factors), for polarization $j$, is then given by (Lock 1987, van de Hulst & Wang 1991)

$$A_j(\lambda,\beta,\theta) = \int_{-\Delta}^{\Delta} S_j(\beta,\theta+\chi)\exp(-i\lambda\chi)\,\mathrm{d}\chi \qquad (15.1)$$

where $S_j$ is the Mie scattering amplitude and

$$\lambda = kr\zeta \qquad (15.2)$$

Since $r\zeta$ represents the distance from the center for a ray in the direction $\zeta$ (fig. 15.1), we see that $\lambda$ represents the *angular momentum* carried by such a ray.

If the lens (observer) is at a very large distance, $\Delta \to 0$ and $A_j$ reduces to $S_j(\beta,\theta)$. At a closer distance, however, we see from (15.1) that $A_j$ is the *Fourier transform of the Mie amplitude over the angular domain covered by the lens*.

If the Mie amplitude within this domain is dominated by WKB contributions from a few Debye terms, its angular dependence will be dominated by travelling-wave Legendre functions of the type

$$Q_{\bar\lambda_p-\frac{1}{2}}^{(1,2)}(\cos\theta)$$

where $\bar{\lambda}_p$ are the corresponding geometrical-optic saddle points in the $\lambda$ plane, and the superscript is 1 or 2 depending on whether the contribution is nearside or farside [cf. (12.8)].

Substituting in (15.1) the asymptotic expansion (7.23) of the Legendre functions, valid away from near-forward or near-backward directions, we see that, for each geometrical-optic saddle point, we find a contribution to $|A_j|$ given by

$$ |A_j|_{\bar{\lambda}_p} \propto \left| \frac{\sin\left[(\lambda \pm \bar{\lambda}_p)\varDelta\right]}{(\lambda \pm \bar{\lambda}_p)} \right| \tag{15.3} $$

Thus, the observed pattern corresponds to a series of peaks at the geometrical-optic ray locations, which are the glare spots. To each of them is associated the diffraction pattern of a slit defined by the lens aperture: enlarging it increases the resolution of the spots, at the expense of decreased resolution in scattering angle. These features are confirmed by numerical evaluations of (15.1) (Lock 1987, van de Hulst & Wang 1991).

In contrast with observations made at very large distances, the contributions from various critical points in the $\lambda$ plane can be well separated in glare spot experiments, thus rendering possible, for instance, the observation of high-order rainbows (Walker 1976, 1977).

In addition to the geometrical-optic (Debye) glare spots, plots of the Fourier transforms of the Mie amplitudes allow one to detect also the sharp ripple resonances. They appear around values of $\lambda$ and $\beta$ corresponding to the resonance condition (14.5) in a given partial wave. In view of the sensitive phase adjustments required for a resonance, the experimental observation of resonance glare spots is more difficult: it has been achieved by employing laser light illumination (Lock & Woodruff 1989).

*Particle size and refractive index determination*

The sensitive dependence of sharp ripple resonance features on the size parameter and the refractive index render them very suitable as probes for high-accuracy determinations of these parameters. As was already mentioned in Sec. 5.2, the resonances can be observed on optically levitated liquid droplets (Ashkin & Dziedzic 1977, 1981, Ashkin 1980).

The largest ratio of resonances to background is achieved in backscattering. In spite of the quasiperiodicity, there are enough distinctive features in resonance shapes and in their changes from one quasiperiod to another to render possible the identification of individual resonances (Eversole *et al.* 1990).

The sharpness of the resonances allows relative and absolute size measurements to be made with an accuracy two to three orders of magnitude higher than that attained by previous light scattering techniques. For relative size measurements, based on wavelength ratios of a given resonance for different spheres, the accuracy reaches one part in $10^5$. Changes in the average diameter of a drop (e.g., due to evaporation) of 1 part in $10^6$, amounting to 0.1 Å for a 10-μm droplet, can be detected (Ashkin 1980).

For a microsphere for which some lower-accuracy a priori knowledge of size and refractive index exists, a simultaneous determination of absolute size and refractive index can be made with an accuracy of about 4 parts in $10^5$, by comparison with the theoretical resonance spectrum (Chylek *et al.* 1983).

Similar accuracies have also been attained in the measurement of the average diameter of low-birefringence optical fibers by resonant light backscattering (Ashkin, Dziedzic & Stolen 1981).

Other recent applications to optical particle sizing are described in Gouesbet & Gréhan 1988. Applications to the spectroscopy of levitated particles and aerosols are described in Barber & Chang 1988.

## 15.2   Applications to radiative transfer and to astronomy

### Radiative transfer

The basic quantity in the theory of radiative transfer (Chandrasekhar 1950) is the *specific intensity of radiation* $I_\nu(r,\hat{s})$ at point $r$ in the direction of the unit vector $\hat{s}$, at frequency $\nu$. This is defined so that the energy $dE_\nu$ crossing a surface element $d\sigma$ perpendicular to $\hat{s}$, during the time interval $dt$, in directions within a solid angle $d\Omega_{\hat{s}}$ around $\hat{s}$, in the frequency range $(\nu, \nu + d\nu)$, is given by

$$dE_\nu = I_\nu(r,\hat{s})\,d\sigma\,d\Omega_{\hat{s}}\,d\nu\,dt \qquad (15.4)$$

The basic *equation of transfer* governing the specific intensity, due to Schwarzschild, is (Chandrasekhar 1950, Goody & Yung 1989)

$$\hat{s}\cdot\nabla I_\nu = \varepsilon_\nu - \alpha_\nu I_\nu \qquad (15.5)$$

which expresses the directional derivative of the specific intensity as the difference between a *source term* and an *extinction term*.

In the extinction term, $\alpha_v$ is called the *volume extinction coefficient*, and it depends on the extinction efficiency $Q_{ext}$ of the particles in the medium. For a scattering medium such as cloud that scatters radiation in the atmosphere, the source term is the atmospheric radiation itself (multiple scattering), and one has

$$\varepsilon = \zeta\,\alpha \int \frac{d\Omega_{\hat{s}'}}{4\pi}\, p(\hat{s},\hat{s}')\,I(r,\hat{s}') \tag{15.6}$$

where we have omitted the indices $v$ for simplicity.

In this equation, $\zeta$ is the *single-scattering albedo*, representing the fraction of the intensity that gets scattered, rather than absorbed:

$$\zeta = \frac{Q_{sca}}{Q_{ext}} = 1 - \frac{Q_{abs}}{Q_{ext}} \tag{15.7}$$

where we have employed (13.3), and $p(\hat{s},\hat{s}')$ is the (unpolarized) *phase function*, proportional to the differential scattering cross section in the direction $\hat{s}'$ for radiation incident in the direction $\hat{s}$. The normalization is such that

$$\int \frac{d\Omega_{\hat{s}'}}{4\pi}\, p(\hat{s},\hat{s}') = 1 \tag{15.8}$$

The phase function is generally assumed to be a function only of the scattering angle $\Theta$:

$$p(\hat{s},\hat{s}') = p(\hat{s}\cdot\hat{s}') \equiv p(\cos\Theta) \tag{15.9}$$

For Mie scattering, the phase function is proportional to $\frac{1}{2}(i_1 + i_2)$, the intensity for scattering of natural (unpolarized) light.

The most important feature of the phase function in radiative transfer problems is the associated asymmetry factor $g$, defined as in (13.4). For practical computations, the Mie phase function is often modelled by the far simpler Henyey–Greenstein phase function (Henyey & Greenstein 1941), which is entirely defined by $g$:

$$p_{\text{HG}}(\cos\Theta, g) \equiv \frac{1 - g^2}{\left(1 + g^2 - 2g\cos\Theta\right)^{3/2}} = \sum_{l=0}^{\infty}(2l+1)g^l P_l(\cos\Theta) \quad (15.10)$$

whose asymmetry factor is $g$.

The above results suffice to justify the special relevance of the three average efficiency factors (13.9) in the applications to radiative transfer [by (13.6), $g$ can be obtained from $Q_{\text{pr}}$ ].

According to (15.6), the equation of transfer (15.5) is an integro-differential equation. Approximate methods for the solution of this equation are discussed in van de Hulst 1980 and Goody & Yung 1989.

### Optical sizing of cometary dust

The 'head' or 'coma' of a comet is the bright, roughly spherical region from which its tail originates. In this region, varying with the distance from the Sun, cometary dust grains with a wide range of sizes are produced, presumably by sublimation. Information about Halley's Comet grain size distribution can be obtained by combining *in situ* sampling results from the 1986 Giotto and Vega space probes with emission spectra from the comet ranging from the visible to the far infrared.

The observed emission continuum results both from the scattering of sunlight and from thermal emission, the latter depending on the absorption efficiency of the grains. The grains are expected to be spinning very fast, so that their average properties may be modelled by taking them to be spherical particles. The input required for the calculation includes the Mie absorption efficiency associated with the grains, for size parameters ranging continuously from very small values to values of order $10^6$. For large size parameters, the CAM expressions discussed in Chapter 13 have been employed (Crifo 1988).

### The Atmosphere of Venus

The observation of Venus, despite its proximity, has been frustrated by the optically thick veil of clouds that surround it. The nature of these clouds has been determined by a beautiful inverse scattering application of Mie theory (Hansen & Arking 1971, Hansen & Hovenier 1974).

The linear polarization of sunlight reflected from the atmosphere of Venus, as a function of the scattering angle, has been measured with great accuracy over the past half century, from the visible to the far infrared.

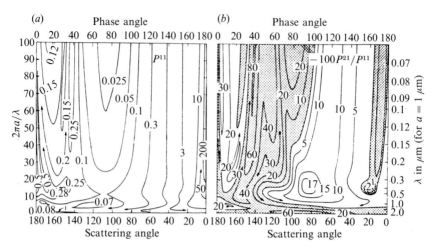

Fig. 15.2. *(a)* Level curves for the phase function of light scattered from a polydisperse distribution of the type (15.11), with $N = 1.33$ and $b = 0.07$; *(b)* Corresponding level curves for the degree of linear polarization (in percent); positive regions are shaded (after Hansen & Travis 1974. Reprinted by permission of Kluwer Academic Publishers).

For a comparison with Mie theory, one must take into account the effect of size dispersion. Hansen & Hovenier (1974) employed a model size distribution function

$$\mathcal{N}(r) = \text{constant} \times r^{(1-3b)/b} \exp[-r/(ab)] \tag{15.11}$$

where $\mathcal{N}(r)\,dr$ denotes the number of particles with radii in the range $(r, r+dr)$. The *effective radius* of this distribution, defined as the weighted average of $r$ with a weight proportional to the geometrical cross section $\pi r^2$, is equal to $a$, and its *effective variance*, defined with the same weight, is equal to $b$.

Level curves for the phase function and the degree of linear polarization (in percent) averaged over the size distribution (15.11), for $N = 1.33$ and $b = 0.07$, as functions of the effective size parameter $ka$ and the scattering angle, are shown in fig. 15.2 (Hansen & Travis 1974).

At the larger size parameters shown, the forward diffraction peak is recognized by the large values of the phase function and absence of polarization. At fairly small scattering angles (around 15° in this figure), the direct reflection contribution is dominant, with positive linear polarization, as follows from the Fresnel reflection coefficients in (9.15) (the parallel component is depressed by the vicinity of the Brewster angle).

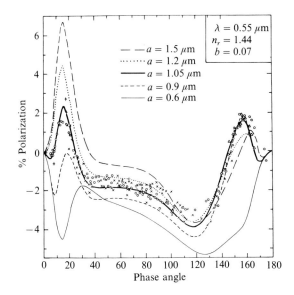

Fig. 15.3. Observations of the polarization of sunlight reflected by Venus in the visible by four different authors (marked points) and theoretical curves for $\lambda = 0.55$ µm, $N = 1.44$, $b = 0.07$ and various values of $a$, including a contribution from Rayleigh scattering. The phase angle is the supplement of the scattering angle (from Hansen & Hovenier 1974. Reprinted by permission of the American Meteorological Society).

At intermediate size parameters in this angular range, one notices the effects of anomalous diffraction (Sec. 9.6). In the remainder of the forward hemisphere, direct transmission predominates, with negative linear polarization [cf. (9.20)].

The peaking of the phase function, accompanied by strong positive polarization, around 150°, corresponds to the primary rainbow; similar but weaker features for the secondary bow are seen around 120°. Finally, the sharp backward peak in the phase function, with rapidly varying polarization around it, is associated with the glory.

We see that the polarization features remain present in spite of the averaging over the size distribution. The effects of multiple scattering for a homogeneous atmosphere have also been taken into account (Hansen & Hovenier 1974), by applying radiative transfer methods with the polarization included.

Fig. 15.3 shows a comparison between the resulting theoretical polarization curves for $N = 1.44$, wavelength $\lambda = 0.55$ µm and $b = 0.07$, for various assumed values of the effective radius $a$, and experimental observations by four different authors. The positive peak at phase angle (supplement of scattering angle) ~ 20° is a primary rainbow (or rather, in view of the size parameter value, fog bow) seen in the clouds of Venus! The other positive peak is due to anomalous diffraction. The best fit in this example occurs for $a = 1.05$ μm.

By combining similar fits in which different parameters are varied and observations at other wavelengths are included, Hansen & Hovenier (1974) finally arrived at the following conclusions: (i) The particles in the visible clouds of Venus are spherical. The observation of rainbows, which are associated with circular cross sections, and the variation of the polarization with scattering angle and wavelength lead to this conclusion. (ii) The refractive index of the particles ranges from $1.46 \pm 0.015$ at $\lambda = 0.365$ μm to $1.43 \pm 0.015$ at $\lambda = 0.99$ μm. (iii) The effective radius of the particle size distribution is $a = 1.05 \pm 0.10$ μm, and its effective variance is $b = 0.07 \pm 0.02$. This is a remarkably uniform distribution by terrestrial standards.

These results allowed them to infer that the cloud particles probably consist of a concentrated solution of sulfuric acid, the only composition, among all those that had been suggested in the literature, that would be consistent with all the data.

Subsequent observations by several space probes have essentially confirmed the validity of all these conclusions for the major constituent of the upper clouds of Venus (Kawabata *et al.* 1980, Knollenberg & Hunten 1980, Marov *et al.* 1980, 1983, Ragent & Blamont 1980, Esposito *et al.* 1983, Krasnopolsky 1985).

## 15.3   Applications to acoustics

The scattering of sound waves by a sphere is a classic problem in acoustics (Rayleigh 1877). For a historical survey of work on this problem, see Brill & Gaunard 1987. Large size parameters in acoustics are chiefly relevant to the scattering of ultrasonic waves or (through Fourier synthesis) sound pulses (Friedlander 1958).

All of the CAM results can be applied to sound scattering. The theory of scalar scattering by an impenetrable sphere treated in Chapters 7 and 8 can be immediately reinterpreted in terms of sound scattering by a perfectly soft sphere, and the results are readily generalized to a perfectly rigid sphere (Nussenzveig & Wiscombe, to be published). Results of the uniform CAM approximation for a rigid acoustic sphere have been displayed in figs. 8.4 and 8.5. The CAM theory for a penetrable fluid sphere, equivalent (for monochromatic waves) to the quantum scattering problem for a rectangular potential, is developed in Nussenzveig 1969a,b).

For an elastic obstacle, where both longitudinal and transverse sound waves can propagate, the CAM theory can also be applied (see the next section). In this case, other types of surface wave modes, including Rayleigh waves and whispering-gallery waves, can propagate (Uberall 1973). Observations of these surface wave modes by schlieren visualization are described in Neubauer 1973.

For a penetrable object, sharp resonances analogous to those treated in Chapter 14 exist. In acoustics, these large-angular momentum resonances are interpreted as whispering-gallery modes (cf. Sec. 14.1). They have been observed in the scattering of ultrasound by an aluminum cylinder in water (Maze, Taconet & Ripoche 1981, Maze & Ripoche 1983). The scattering of each individual resonant partial wave has also been described theoretically in terms of $S$-matrix pole contributions and background contributions associated with scattering by an impenetrable sphere (Flax, Dragonette & Uberall 1978, Flax, Gaunard & Uberall 1981).

Acoustic backward glory scattering associated with real glory rays has been observed in the scattering of ultrasound by an elastic sphere in water (Marston, Williams & Hanson 1983b, Williams & Marston 1985a,b). A characteristic glory angular distribution [cf. (4.1)] was found.

## 15.4  Applications to seismology

Seismic body waves, which propagate through the Earth's interior, are both transverse or shear waves, known as $S$ waves, and longitudinal waves, known as $P$ waves (these initials come from 'secondary' and 'primary', referring to arrival times in seismograms: longitudinal waves travel faster). The $S$ waves are further classified into horizontal ($SH$) and vertical ($SV$) polarization components (Bath 1979).

Seismic signals are pulses, with a Fourier spectrum of relevant wavelengths for body waves ranging up to at most a few hundred kilometers, so that the Earth's size parameter is typically of order $10^2$ or greater. Under these conditions, semiclassical diffraction effects play a significant role in seismic wave propagation.

Early applications of the Watson transformation to spherical Earth models were made by Scholte (1956) and by Duwalo & Jacobs (1959). The problem with which they were concerned was that of the diffraction of *SH* waves by the Earth's liquid core, within which shear waves cannot propagate, so that this problem is analogous to that of an impenetrable sphere. For simplicity, one can first neglect the effect of the Earth's free surface, so that the Earth's mantle is treated as an unbounded homogeneous elastic medium.

What must be treated in seismology is the diffraction of waves from a point source, so that the starting point is the exact partial-wave expansion of Green's function. The CAM treatment given in Chapter 7 has been extended to this situation by Ansell (1970, 1978). The Fock approximation in the penumbra region and the shift of the shadow boundary in the Fresnel region (cf. Sec. 7.5) were obtained. Applications of this shift to seismology were discussed by Teng & Richards (1969).

The propagation of *P* waves, that can travel through the core, corresponds to scattering by a penetrable sphere. One complication is that, at a solid-fluid boundary, there is mode conversion: a *P* wave is reflected partially as *P* and partially as *SV*, with different angles of reflection, corresponding to the different velocities of propagation.

When the free surface of the Earth is taken into account, the Debye expansion must be generalized to a layered medium, as a double Debye expansion, or as a multiple one (Gerard 1979, 1980) if one takes into account also the existence of the Earth's inner core, which is assumed to be solid.

The Debye expansion plays a central role in seismology, as it is the basis for the notation employed to characterize different seismic signals that are observed, associated with different travel times. This is illustrated in fig. 15.4 for a *P* wave point source. When the direct *P* waves reach the core, they are reflected both as *S* waves (*PcS*) and as *P* waves (*PcP*), the notation *c* being employed to denote reflection from the core boundary. Within the core, the refracted wave is denoted by *K* (from the German 'Kern' for 'core'), and direct transmission terms are therefore denoted by *PKS* and *PKP*. The direct *P* wave is also reflected from the Earth's surface, giving rise to the wave *pP* in fig. 15.4, and similarly for higher-order Debye terms.

Reviews of the applications of CAM and other methods in seismology are given by Chapman & Phinney 1972, Aki & Richards 1980, Ben-Menahem & Singh 1981 and Hanyga 1985.

Diffraction effects relevant to seismology include creeping-wave penetration into the shadow of the core (Chapman & Phinney 1972, Doornbos & Mondt 1979a,b), diffraction near a caustic analogous to that of the primary rainbow for *PKKP* waves (Richards 1973) and tunneling

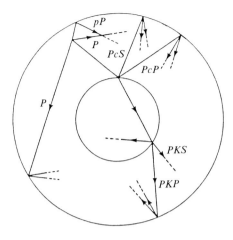

Fig. 15.4. Illustration of some low-order Debye terms generated by a $P$ wave point source (from Chapman & Phinney 1972. Reprinted by permission of Academic Press).

into the core from mantle ray paths passing outside of it (Engdahl 1968, Adams 1972, Richards 1973). For interfaces with media of lower refractive index, involving total reflection (such as occur at the inner-core/outer-core boundary), near-critical scattering effects similar to those discussed in Chapter 12 take place, giving rise to seismic head waves (Cerveny & Ravindra 1971) and to whispering-gallery modes (Cormier & Richards 1977, Aki & Richards 1980).

## 15.5   Nonlinear Mie scattering

As was mentioned in Sec. 14.1 and is illustrated in fig. 14.2, the ratio of internal to external field amplitudes in a transparent sphere can take on very large values at a narrow resonance. This can give rise to nonlinear optical effects, some of which were already mentioned in Sec. 5.3.

Numerical studies and experimental observations of internal and near-field intensities for nonresonant and resonant conditions have been made by Chylek, Pendleton & Pinnick (1985), Benincasa *et al.* (1987), and Barton, Alexander & Schaub (1988, 1989). They confirm the qualitative picture given in Chapter 14: on or off resonance, there is an axial focusing effect that produces strong forward peaking as well as backward peaking within the sphere. At resonance, there is a sharp enhancement of the internal intensity, peaking radially in a layer next to the surface; the outside intensity decreases very sharply as one moves away from the surface. Coupling to the resonance is improved when the incident beam is focused close to the edge (Baer 1987, Zhang, Leach & Chang 1988).

Regarding the sphere as an optical cavity, the $Q$ factor associated with a resonance is given by the ratio of the resonance position to its halfwidth. The availability of large $Q$ factors has allowed the observation of lasing from liquid droplets doped with a dye (Tzeng *et al.* 1984, Qian *et al.* 1986, Lin *et al.* 1986), as well as from a solid (Nd:YAG) sphere (Baer 1987). The laser emission is confined to the neighborhood of the droplet boundary, as shown by the photographs in Qian *et al.* 1986.

Stimulated Raman scattering (SRS) from microdroplets has been observed (Snow, Qian & Chang 1985). The density of resonances is high enough for one or more of them to fall within the Raman gain profile, and first-order SRS builds up from spontaneous and amplified Raman scattering within the focal region. The resonant field intensity enhancement around the droplet from the first-order Stokes emission is so strong that it can act as a pump for second-order Stokes emission, and so successively: up to 14th-order sequentially pumped SRS from $CCl_4$ microdroplets has been detected (Qian & Chang 1986). Studies of the time dependence of these processes agree with the proposed temporal sequence and have led to measured $Q$ factors in SRS exceeding $2 \times 10^7$ (Zhang, Leach & Chang 1988, Hsieh, Zheng & Chang 1988, Pinnick *et al.* 1988). The angular distribution of SRS has been observed and employed to identify the specific Mie resonances involved (Chen *et al.* 1991).

Optical third-harmonic generation, and other third-order sum-frequency generation in microdroplets, enhanced by the Mie resonances, have been reported (Acker, Leach & Chang 1989). Stimulated Brillouin scattering has also been observed (Zhang & Chang 1989) and theoretically approached (Chitanvis & Cantrell 1989, Ching, Leung & Young 1990).

The microdroplet resonance $Q$ factors observed in elastic scattering, even in the linear regime, have been limited to about $10^6$. In order to account for this limitation, which does not arise from material absorption, it has been proposed that it may originate from shape distortions associated with capillary waves excited by thermal fluctuations (Lai *et al.* 1990, Lai, Leung & Young 1990). This would break the $(2l+1)$-fold degeneracy of a multiplet corresponding to a resonance in the $l$th partial wave, leading to the observed $Q$ reduction in processes that couple to the multiplet as a whole, but it would still allow, in principle, the coupling to a single component with much higher $Q$.

From the viewpoint of fundamental laser theory, the sharp Mie resonances are of considerable interest as a set of laser modes (Nussenzveig 1973, Baseia & Nussenzveig 1984) with accurately determined three-dimensional structure, quite different from the usually employed quasi-one-dimensional models. The very high resonance field gradients near the surface might be of interest in connection with trapping (Nussenzveig 1989a).

The theoretical treatment of nonlinear Mie scattering requires a reformulation of the Mie model, taking into account self-consistently the mutual effects of the coupling between the field and the material properties. Some theoretical discussions have been given (Kurizki & Nitzan 1988, Lai, Leung & Young 1988). It would be specially interesting to extend to the visible domain the results already verified with microwaves concerning quantum-electrodynamic cavity effects (Haroche 1984).

Cavity-QED-enhanced stimulated emission in the visible has recently been observed in dye-doped ethanol droplets for $Q$ values as low as $10^3$–$10^4$ (Campillo, Eversole & Lin 1991).

# Applications to atomic, nuclear and particle physics

*Do we have enough imagination to see in the spectral curves the same beauty we see when we look directly at the rainbow?*
(Feynman, Leighton & Sands 1964)

We now return to the context of particle scattering discussed in Chapter 1, to illustrate applications of semiclassical diffraction effects in quantum-mechanical problems.

(i) In atomic and molecular scattering, the effects already observed include diffractive and rainbow scattering, glories, and orbiting and shape resonances.

(ii) In nuclear heavy-ion scattering, they include nuclear rainbows, backward and forward nuclear glories, and surface waves in diffractive scattering.

(iii) Finally, in particle physics, a phenomenological model based on surface-wave tunneling gives a very good fit of high-energy proton-proton scattering.

## 16.1 Atomic diffractive and rainbow scattering

The typical shape of the potential describing the interaction between two neutral atoms is sketched in the inset of fig. 1.2(a): a repulsive central core, arising from the overlap of the electron clouds, surrounded by an attractive long-range interaction of van der Waals type. An often used model of this type is the Lennard-Jones (12,6) potential

$$V(r) = \varepsilon\left[(r_m/r)^{12} - 2(r_m/r)^6\right] \tag{16.1}$$

where $\varepsilon$ is the well depth, at the position $r_m$ of the potential minimum, which is a measure of the radius of the potential.

The near-forward diffractive scattering by a potential of this type, having a long-range tail, has a very different origin from that treated in Chapter 8: instead of edge diffraction, one must consider the contributions from very large angular momentum paths that undergo small deflections due to the tail of the potential.

A uniform approximation for the forward diffraction peak based on the Poisson representation has been obtained by Berry (1969). As shown in fig. 1.2, one also gets a real forward glory path for such a potential, related with the coalescence of paths such as 2 and 4 in fig. 1.2(b). A uniform expression for the forward glory contribution, based on the CFU method (cf. Sec. 10.3), with axial focusing taken into account, was also derived by Berry (1969). Both approximations have been compared with numerical partial-wave results for several potential models and have been found to be very accurate under semiclassical conditions (Mount 1973).

CAM representations of the scattering amplitude have been derived by several authors (Bosanac 1978a, Connor 1980, Thylwe 1983, Thylwe & Connor 1985). Regge poles and trajectories for various interatomic potentials have been computed (Connor, Delos & Carlson 1976, Bosanac 1978c).

As illustrated in fig. 1.2, potentials of this type also give rise to rainbow scattering. The transitional (Airy) approximation to the rainbow scattering amplitude was given by Ford & Wheeler (1959). A uniform approximation of CFU type was derived by Berry (1966), who verified its accuracy for a Lennard-Jones potential (cf. also Connor & Marcus 1971, Mullen & Thomas 1973, Connor, Farrelly & Mackay 1981).

The first observation of an atomic rainbow was made by Beck (1962), in the scattering of K from Kr and HBr. Fig. 16.1 shows the differential cross section for the scattering of Na from Hg at five different energies (Buck & Pauly 1971). The main rainbow peak and several supernumeraries are detected. The additional superimposed oscillations, seen most clearly at the energies of 0.188 eV and 0.251 eV, and known as 'rapid quantum oscillations', arise from interference with repulsive paths like path 2 in fig. 1.2(a). They are the analogue of the rapid oscillations in fig. 10.6, that are due to interference with direct reflection.

The shift of the rainbow angle with the energy in fig. 16.1 is due to the energy dependence (dispersion) of the 'refractive index' (1.9), so that the results plotted are like spectral curves for different colors; since the wave number also changes with the energy, the size parameters are varying as well.

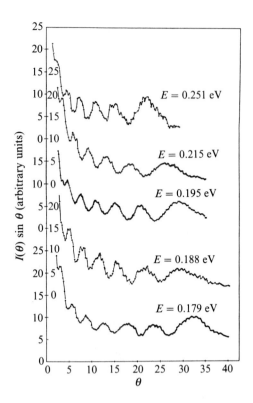

Fig. 16.1. Measured differential cross sections in the center of mass system (in arbitrary units) for the scattering of Na on Hg at five different energies. The curves are arbitrarily shifted from each other (from Buck & Pauly 1971).

The position of the (geometrical-optic) rainbow angle depends only on the refractive index $N$. Correspondingly [cf. (1.9)], in atomic scattering, it yields information on the *depth* $\varepsilon$ of the attractive well in the interatomic potential.

The supernumerary rainbows, as diffraction features, depend on the size parameter, so that their positions give information on the *range* of the interatomic potential. The size-dependence also appears in the rapid quantum oscillations. Since the supernumeraries arise from interference between two attractive trajectories, while the rapid oscillations arise from interference between attractive and repulsive trajectories (fig. 1.2), they probe different parts of the potential, so that, by combining them, one can get information about the *shape* of the interatomic potential (Buck 1974, 1975, 1988).

Rainbow effects have also been detected in rotational and

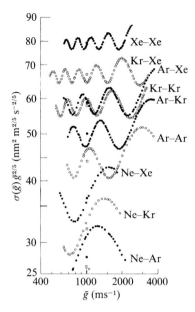

Fig. 16.2. Glory undulations in the total cross section as a function of the relative velocity for several noble gas systems (from van den Biesen 1988).

vibrational excitations, in the scattering of atoms and molecules by surfaces, and in ion channeling. For reviews and further references, see Kleyn 1987 and Neskovic 1990.

## 16.2 Atomic glories and orbiting resonances

As was mentioned in Sec. 16.1, typical interatomic potentials give rise to real forward glory paths, by compensation of positive and negative deflections between the attractive and repulsive regions of the potential traversed by a path. The interference between such paths and large impact parameter forward-diffracted trajectories gives rise to forward glory oscillations in the total cross section that are the real-glory-ray analogues of the forward optical glory oscillations discussed in Sec. 13.4.

Glory undulations in the total cross section observed in several noble gas systems (van den Biesen, Hermans & van den Meijdenberg 1982) are shown in fig. 16.2.

What determines the spacing of the glory undulations? By analogy with the interference between diffraction and direct transmission in anomalous Mie diffraction (Sec. 9.6), we can estimate the phase difference

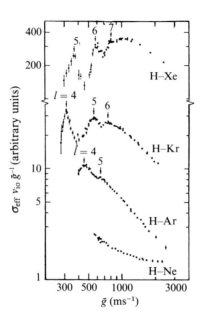

Fig. 16.3. Measured effective cross sections of various hydrogen–noble gas systems. Orbiting resonances are indicated by arrows, together with the corresponding $l$ values (from van den Biesen 1988).

between the two interfering contributions as (Kong, Mason & Munn 1970)

$$\Delta = (N-1)kd \sim \frac{\varepsilon}{2E}kd \propto \frac{\varepsilon r_m}{\hbar v} \tag{16.2}$$

where $d$ is the effective length of the potential slice traversed by the path, $\varepsilon$ and $r_m$ are the strength and range of the potential, respectively [cf. (16.1)], and $v$ is the relative velocity; we have also assumed that $\varepsilon/E$ is small to approximate the refractive index (1.9). We see, therefore, that the spacing of the glory undulations measures the strength-range product $\varepsilon r_m$ (Bernstein 1966).

The total number of glory oscillations that are observed also has a physical interpretation (Bernstein 1966, Kong, Mason & Munn 1970). Since the phase difference between successive interference maxima is $\pi$, and the phase difference approaches the $s$-wave phase shift as $E \to 0$, the total number of oscillations is the same (under semiclassical conditions) as the number of multiples of $\pi$ contained in the zero-energy phase shift. By Levinson's theorem (Levinson 1949, Nussenzveig 1972a), this number is equal to the number of bound states in the potential, i.e., for the present system, the number of vibrational states for the di-atom. Improved estimates of the glory undulations, based upon a uniform approximation, are given in Mason *et al.* 1982.

When the centrifugal barrier is added to the interatomic potential, the resulting effective potential resembles those sketched in fig. 1.3, leading to the possibility of orbiting (shape) resonances at low energies. The first clear evidence for their presence was obtained by Schutte *et al.* (1972). Fig. 16.3 shows the results of Toennies, Welz & Wolf (1979) for the scattering of H atoms by noble gases. The rotational quantum numbers *l* of the resonating quasibound states are indicated.

A CAM treatment of the orbiting resonance contributions to the total cross section was given by Bosanac (1978b). As was found in Chapter 14, they correspond to the contributions from Regge trajectories, with the background given in this case by diffraction, direct reflection and glory contributions. The effects of orbiting resonances on differential cross sections have also been treated by CAM methods (Bosanac 1979, Korsch & Thylwe 1983). Orbiting resonances play an important role in rotational predissociation, in recombination reactions and in several other processes (for references, see Toennies, Welz & Wolf 1979).

The main object of atomic and molecular scattering experiments is the determination of the interactions between the colliding particles. The connections between several observed features of semiclassical diffraction effects and corresponding features of the interaction potential have been pointed out above. By combining these observations with other available data, one can apply inversion techniques to reconstruct the potential (Buck 1974, 1975, 1988). Thus, for instance, it has been found that the Li–Hg potential differs considerably from the Lennard-Jones (12,6) model: it has a much softer repulsive core and a faster long-range decay.

## 16.3.  Rainbows in nuclear physics

We refer to Nörenberg & Weidenmüller (1976) for an introduction to nuclear heavy-ion collisions, and to Brink (1985) for the applications of semiclassical methods in this domain.

Two important differences between nuclear heavy-ion collisions and the atomic collision situations discussed above are the presence of the Coulomb interaction and the existence of strong absorption for collisions with appreciable nuclear overlap, i.e., at distances below a *strong absorption radius* $R_s$. We deal only with elastic heavy-ion scattering.

One can also define a characteristic Coulomb length parameter $a_C$ for a collision with center-of-mass energy $E$ between two nuclei with charge numbers $Z_1$ and $Z_2$,

$$a_C = Z_1 Z_2 e^2 / (2E) \qquad (16.3)$$

Besides the 'nuclear' size parameter $kR_s$, there is also a Coulomb one, the *Sommerfeld parameter* $n = ka_C$. For collisions with laboratory energy exceeding about 10 MeV per nucleon, unless the two partners are light, both parameters are large, so that semiclassical approximations are applicable.

The nuclear interaction is usually described by an optical potential, a common model being the Woods–Saxon potential

$$V(r) = -\frac{V_0}{1 + \exp\left[(r - R_r)/a_r\right]} - \frac{iW_0}{1 + \exp\left[(r - R_i)/a_i\right]} \qquad (16.4)$$

where the imaginary part describes absorption from the elastic channel. The real part corresponds to a rounded-wall attractive well with depth of the order of $V_0$, range of the order of $R_r$, and surface diffuseness $a_r$, with corresponding interpretations for the imaginary part. When $R_i < R_r$, one has a *surface transparent potential*.

The effective potential for a given angular momentum is the sum of the repulsive Coulomb potential, which dominates at large distances, with the attractive (and absorptive) nuclear optical potential and the centrifugal barrier. Typical effective potential shapes for the real part are illustrated in fig. 1.3.

For small Sommerfeld parameter, the Coulomb interaction is less effective, and the elastic scattering, as a first approximation, resembles that from an impenetrable sphere (corresponding to the strong nuclear absorption), showing an Airy-like near-forward diffraction pattern, with the fringe contrast reduced by Coulomb and surface diffuseness effects. This is known as 'Fraunhofer scattering' (Brink 1985).

For larger values of the Sommerfeld parameter, let us begin by considering the qualitative behavior of the deflection function in the absence of nuclear absorption. As the impact parameter is decreased from very large values, only the Coulomb repulsion would be felt at first, leading to positive and growing deflection angles, until the effects of nuclear attraction begin to bend the trajectories back. This would lead to a maximum in the deflection function, corresponding to the *Coulomb rainbow*.

With further decrease of the impact parameter, one would (still neglecting absorption) penetrate into the region where the effective potential is dominated by the nuclear attraction and the centrifugal barrier, and the qualitative behavior of the deflection function would be similar to that shown in fig. 1.2: it would go through a minimum, corresponding to a *nuclear rainbow*.

The effect of nuclear absorption is to damp out the contributions from paths located below the strong absorption radius. For weak enough absorption, one may recognize a Coulomb rainbow, that can be well fitted by the Berry (1966) uniform approximation (da Silveira 1973). Uniform rainbow approximations including absorption have also been derived (da Silveira 1990, Pato & Hussein 1990). The effect of the Coulomb repulsion on the paths simulates that of a divergent lens with a virtual focus at finite distance, leading, for strong nuclear absorption, to a Fresnel-like diffraction pattern (Frahn 1985). There is some ambiguity as well as overlap between the Coulomb rainbow and Fresnel diffraction descriptions (Brink 1985).

If the absorption is not very strong, the nuclear rainbow may also become observable. To have a clear signature, it is convenient, in a partial-wave expansion based upon an optical-model fit to data, to apply the traveling-wave Legendre functions of (7.22) in order to obtain a nearside–farside decomposition (cf. Sec. 12.2). In nuclear physics, this approach was introduced by Fuller (1975) and it has been extensively reviewed by Hussein & McVoy (1984). Nearside paths are repulsive, so that the Coulomb rainbow would be a nearside feature; farside paths are attractive: thus, one should look for a nuclear rainbow in the farside amplitude.

Fig. 16.4 shows a logarithmic plot of the differential cross section for the scattering of $\alpha$ particles by $^{40}$Ca at 50 MeV, measured by Delbar *et al.* (1978) and the corresponding nearside–farside decomposition (McVoy *et al.* 1986, McVoy 1990). The farside amplitude shows a clear-cut nuclear rainbow, with the primary bow peaking around 140° and the first supernumerary quite apparent in the cross section, peaking around 75°. The oscillations around 30° arise from interference with the farside amplitude, and they may be compared with the interference with direct reflection in fig. 10.6.

Another example of a nuclear rainbow occurs in the scattering of $\alpha$

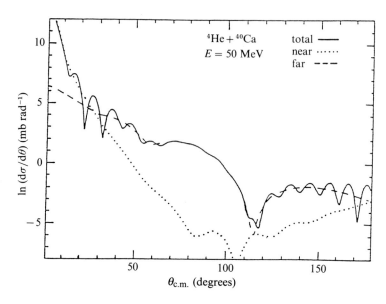

Fig. 16.4. Logarithmic plot of the differential cross section for the scattering of α particles by $^{40}$Ca, measured by Delbar *et al.* (1978), and corresponding nearside–farside decomposition (after McVoy *et al.* 1986).

particles by $^{90}$Zr (Put & Paans 1977, Hussein & McVoy 1984). In this case, the falloff on the shadow side is clearly visible at the highest energies measured.

## 16.4   Nuclear glories and surface waves

*Backward glory and anomalous large-angle scattering*
A sample of experimental data on α particle near-backward scattering from some light spinless nuclei is shown in fig. 16.5. These and other similar data can be reasonably well fitted by an angular distribution of the glory type, given by (4.1), as illustrated in the figure (Bryant & Jarmie 1968).

The glory angular momentum in these fits is close to a grazing angular momentum such that the corresponding Coulomb trajectory is just impinging on the strong absorption radius. In most cases, the dominant contribution to the glory comes from surface waves (Brink 1985).

An extension of CAM theory to heavy-ion nuclear scattering,

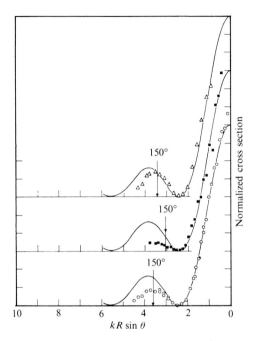

Fig. 16.5. Normalized differential cross section for α particle elastic near-backward scattering from some spinless nuclei: $^{16}O$ at 18.3 MeV (upper curve); $^{12}C$ at 21.9 MeV (middle curve); $^{12}C$ at 18.5 MeV (lower curve). The abscissa is $kR\sin\theta$, where $R$ represents the glory impact parameter, and the full-line curves are fits to a $J_0^2$ angular distribution (from Bryant & Jarmie 1968).

taking into account the Coulomb interaction, was given by Fuller & Moffa (1977). In this treatment, Regge pole contributions are interpreted in terms of surface waves and diffracted rays similar to those of fig. 4.3, except for the curvature arising from the Coulomb potential. Good fits to diffractive oscillations around the Coulomb rainbow region were obtained by a parametrization in which the nuclear contribution is described by a single Regge pole.

For a surface-transparent potential, trajectories can penetrate within the region where they are affected by the real part of the nuclear potential before strong nuclear absorption sets in. One may then decompose the amplitude into a direct reflection term (barrier wave), corresponding to reflection at the Coulomb barrier, and a remainder that feels the effects of the nuclear potential (internal wave). This decomposition has been employed by Brink & Takigawa (1977). It was applied to the backward nuclear glory by Takigawa & Lee (1977) (cf. also Lee 1978). The internal wave contribution is treated by the uniform glory approximation (Berry 1969).

In some light systems, e.g., in elastic α particle scattering from calcium isotopes, the backward glory is accompanied by enhanced scattering at large angles, with complicated angular distributions (anomalous large angle scattering). These effects have been interpreted in terms of interference between the barrier and internal amplitudes (Brink 1985).

One can also apply a Debye-type expansion to the internal amplitude (Agassi & Avishai 1978, Anni, Renna & Taffara 1978, 1980a,b, Anni & Renna 1981, Di Salvo & Vianno 1977) and fit the results for the backward glory and anomalous large-angle scattering in terms of surface waves associated with the Regge–Debye poles, which, in this connection, have been called 'Sommerfeld poles' (Di Salvo & Vianno 1982, Di Salvo 1983).

### Nuclear forward glory

The observability of forward glories in optics as well as in atomic scattering led to the suggestion (Hussein *et al.* 1982) that they might also be observable in heavy-ion nuclear scattering. While in the optical case one must take into account the interference with forward diffraction (Sec. 13.4), the nuclear problem is complicated by the interference with Coulomb scattering, with its characteristically singular amplitude (Taylor 1974) as the forward direction is approached.

In order to extend the optical theorem to this situation, one considers the 'sum-of-differences' cross section

$$\sigma_{\text{sod}}(\theta_0) \equiv 2\pi \int_{\theta_0}^{\pi} \left( \frac{d\sigma_C}{d\Omega} - \frac{d\sigma_{\text{el}}}{d\Omega} \right) \sin\theta \, d\theta \qquad (16.5)$$

where the expression within parentheses is the difference between the Coulomb and elastic differential cross sections, and the integration extends down to a small angle $\theta_0$. It can then be shown (Marty 1983, Barrette & Alamanos 1985, Lipperheide 1987) that, for small enough $\theta_0$,

$$\sigma_{\text{sod}}(\theta_0) \approx \sigma_R - \frac{4\pi}{k} |f_N(0)| \sin\left[ \arg f_N(0) - 2\sigma_0 + 2n \ln \sin\frac{\theta_0}{2} \right] \qquad (16.6)$$

where $\sigma_R$ is the total reaction cross section, $f_N(0)$ is the forward nuclear scattering amplitude, $\sigma_0$ is the $s$-wave Coulomb phase shift and $n$ is the Sommerfeld parameter. This result is a charged-particle analog of the optical theorem (7.6), expressing the removal of flux from the Coulomb trajectories by absorption and elastic scattering in terms of a 'shadow' relative to the Coulomb wave within a narrow forward cone (Lipperheide 1987).

The singular behavior as $\theta_0 \to 0$ is apparent from (16.6): the last term on the right-hand side oscillates with constant amplitude but increasingly fast frequency. The existence of such angular oscillations with an amplitude comparable to $\sigma_R$ would be a signature of the presence of a nuclear forward glory and would allow one to measure $\sigma_R$ in a model-independent way. They would also provide information about the magnitude and phase of the nuclear forward scattering amplitude and, in view of the sensitivity of a forward glory trajectory to the nuclear interaction at short distances, they could allow one to discriminate among different nuclear models. The first term within square brackets in (16.6), originating from the glory trajectory, should also give rise to glory undulations as a function of energy, with a local period inversely proportional to the glory collision time (Hussein *et al.* 1982).

According to (16.6), the detection of a nuclear forward glory is a difficult experiment, requiring high-precision measurements at very small angles. It was successfully performed by Ostrowski *et al.* (1989), who observed it in $^{12}C + {}^{12}C$ scattering at $E_{c.m.} = 9.5$ MeV. Their results are shown in fig. 16.6, together with the theoretical fit (the Coulomb cross section must be replaced by the Mott cross section for identical particles). The oscillatory dependence on angle is clearly present and leads to a determination of the total reaction cross section.

The corresponding results for $|f_N|^2$, shown in the inset of fig. 16.6, are well fitted by $J_0^2(l_G \sin\theta)$ with $l_G = 6$, the value associated with the grazing angular momentum in this collision; the amplitude of the oscillations is comparable with $\sigma_R$, so that the glory signature is well established. An optical model calculation, also represented in the inset, does not exhibit a nuclear forward glory, and comparisons with values of $f_N(0)$ obtained from optical model potentials that yield good fits to the angular distributions in $^{12}C + {}^{12}C$ scattering show that none of them reproduces the correct amplitudes (Ostrowski *et al.* 1989). Thus, the information supplied by the detection of the nuclear forward glory does lead to new constraints on nuclear models.

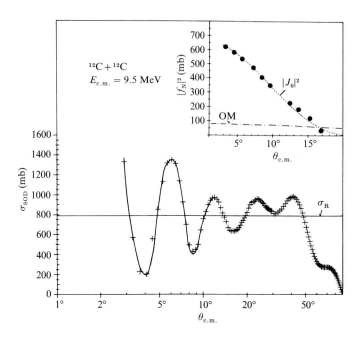

Fig. 16.6.  Sum-of-differences cross section (16.5) for $^{12}C + {}^{12}C$ at $E_{c.m.} = 9.5$ MeV; the crosses are obtained from experiment and the solid line is the theoretical fit. The inset shows the resulting values of $|f_N|^2$ and a fit by $J_0^2(6\sin\theta)$ (dotted line), compared with the result of an optical model calculation (dash-dotted line) (from Ostrowski *et al.* 1989).

### Surface waves in high-energy nuclear physics

Angular distributions for proton scattering at a laboratory energy of 1.98 GeV from nuclei ranging over the periodic table, from $^{12}C$ to $^{208}Pb$, show an Airy-like near-forward diffraction pattern corresponding to a size parameter $kR_A$ (where $R_A$ is the nuclear radius) which ranges from about 20 to 60. If the differential cross sections are plotted against the Airy variable $kR_A\theta$, their falloff at larger angles is on the average exponential and shows strong antishrinkage as the size parameter increases.

It is quite remarkable (Schrempp & Schrempp 1980) that, when the same data are plotted against the variable $(kR_A)^{1/3}\theta$ which is associated with surface-wave decay [cf. (7.9)], the slopes in a logarithmic plot all become parallel. Furthermore, as shown in fig. 16.7, when one normalizes the differential cross sections in such a plot, by dividing them by $R_A^2$, one finds evidence of universality in behavior throughout the periodic system. This zero-parameter fit strongly suggests that the dynamics of diffraction within the 'nuclear shadow', under semiclassical conditions, is governed by surface waves.

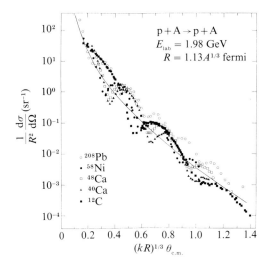

Fig. 16.7. Differential cross sections for proton elastic scattering off nuclei throughout the periodic table, at 1.98 GeV laboratory energy, divided by $R_A{}^2$ ($R_A$ = nuclear radius), plotted against $(kR_A)^{1/3}\theta$, fall on a universal curve. The solid line is a fit to a representation by a sum of surface waves (from Schrempp & Schrempp 1980).

## 16.5  Application to particle physics

Several diffractive models of high-energy nucleon–nucleon scattering have been formulated (Alberi & Goggi 1981, Kamran 1984). Backward peaks have been observed, particularly in pion–nucleon scattering, for which backward glory interpretations have been proposed (Di Salvo & Vianno 1977, Lyubimov 1978). We describe a model due to Schrempp & Schrempp (1977a,b, 1980), in which hadronic diffraction is described as a tunneling phenomenon associated with surface waves.

The basic idea is that, in high-energy elastic collisions, hadrons can be treated as extended objects, in the sense that they interact during the collision over an extended interaction region with a characteristic shape (in the center-of-mass system): a constant transverse radius $R_T$, of the order of 1 fm, and a longitudinal radius $R_L$ that grows linearly with the momentum $k$, so that

$$R_L \approx ckR_T{}^2 \tag{16.7}$$

where $c$ is a constant. Thus, the interaction region at high energies

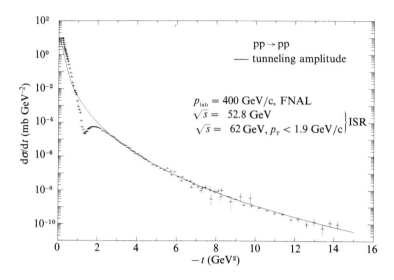

Fig. 16.8. Comparison between experimental data on p–p scattering at high energies (crosses) and tunneling amplitude with parameters $R_T = 1.1$ fm, $c = 0.086$ and an additional surface diffuseness parameter (from Schrempp & Schrempp 1980).

becomes a prolate spheroid, and $c$ can be taken as a measure of its eccentricity. For a dynamical justification of this model in terms of color electric flux tubes or bag models, we refer to Schrempp & Schrempp (1980).

It is also assumed that hadron interpenetration leads to particle production and total loss from the elastic channel, so that contributions to elastic scattering arise only from impact parameters above $R_T$. The collision is then treated in terms of an analogue model, the scattering of a plane wave with wave number $k$ by a ($k$-dependent) impenetrable prolate spheroid with axes related by (16.7).

The scattering amplitude can be expressed in terms of a background integral plus a residue series arising from Regge-like poles, that represent surface waves travelling along the surface of the spheroid and are associated with tunneling (Levy & Keller 1960). Near-forward scattering gives rise to an Airy-like pattern governed by $kR_T$, while large-angle diffraction becomes dominated by surface-wave damping. Schrempp & Schrempp (1980) also treat a model with one extra parameter, to represent blurring of the sharp boundary of the interaction region.

Fig. 16.8 shows their fit from this model to proton–proton scattering data at very high energies. Only the surface-wave contribution is taken into account, so that the Airy-like diffraction dip is not reproduced. The fit is good over two regions of the transverse momentum $p_T$ with quite different behavior, covering a range of several orders of magnitude: (i) the intermediate region where one finds an Orear-type (Orear 1964) exponential behavior $\exp(-Ap_T)$; (ii) the large-$p_T$ region where one gets an inverse power law behavior.

The behavior in region (ii), which has been interpreted as arising from the presence of pointlike constituents of hadrons, arises here from scattering by the pointlike tip of the spheroid.

## 16.6  Why complex angular momentum?

All of the semiclassical critical effects classified in Chapter 1 (as well as a new one, near-critical scattering) are found in Mie scattering. As we have seen, CAM theory leads to very accurate approximations for the associated diffraction phenomena.

We have briefly discussed macroscopic and microscopic manifestations of these phenomena in a great variety of fields. In many cases, CAM methods have also been applied.

One may well ask: Why is CAM theory so effective? Is there not some other (possibly simpler or more intuitive) approach that would yield similar results? To some extent, of course, simplicity is in the eye of the beholder, so that the last question is not very well posed. However, one may provide some arguments to justify the use of CAM theory.

We have found that tunneling plays a prominent role in all semiclassical diffraction phenomena. The main features that were missing in prior treatments are connected with tunneling.

Do we have an intuitive physical picture of tunneling? As an aspect of wave propagation, it may be regarded as a classical effect, already known to (and discovered by) Newton in frustrated total reflection. However, all descriptions of tunneling make use, in one form or another, of analytic continuation. Ultimately, this goes back to Euler's great discovery of the connection between oscillations and the exponential function, exemplified by what has been called one of the most beautiful formulas of mathematics, his synthesis of analysis, algebra, geometry and arithmetic: $e^{i\pi} = -1$.

Though alternative methods of analytic continuation have been applied, CAM theory is especially well suited to the treatment of semiclassical scattering problems: angular momentum is conjugate to the scattering angle, and the localization principle provides a natural physical interpretation of the Poisson representation, linking Huygens' principle with pseudoclassical paths. This allows one to combine physical insights derived from classical wave theory and from quantum potential scattering in order to deal with previously intractable dynamical features of diffraction.

Thus, the beautiful mathematical theory of analytic continuation provides the key to a deeper understanding of some of the most beautiful phenomena displayed in the sky, and also manifested in so many other ways – through all scales of size – revealing the underlying unity of nature.

# References

Abramowitz, M. & Stegun, I. A. (1965) *Handbook of Mathematical Functions*. New York: Dover.

Acker, W. P., Leach, D. H. & Chang, R. P. (1989) *Opt. Lett.* **14**, 402.

Adams, R. D. (1972) *Bull. seism. Soc. Am.* **62**, 1063.

Agassi, D. & Avishai, Y. (1978) *Phys. Lett.* B**74**, 18.

Aki, K. & Richards, P. G. (1980) *Quantitative Seismology*, vol. 1. San Francisco: W. H. Freeman.

Alberi, G. & Goggi, G. (1981) *Phys. Rep.* **74**, 1.

Anni, R., Renna, L. & Taffara, L. (1978) *Nuovo Cimento* A**45**, 123.

Anni, R., Renna, L. & Taffara, L. (1980a) *Nuovo Cimento* A**55**, 456.

Anni, R., Renna, L. & Taffara, L. (1980b) *Nuovo Cimento* A**59**, 38.

Anni, R. & Renna, L. (1981) *Nuovo Cimento* A**65**, 311.

Ansell, J. H. (1970) Ph. D. thesis, Cambridge University.

Ansell, J. H. (1978) *Geophys. J. R. astr. Soc.* **54**, 349.

Arnold, V. I. (1975) *Russ. Math. Survs.* **30**, 1.

Ashkin, A. & Dziedzic, J. M. (1977) *Phys. Rev. Lett.* **38**, 1351.

Ashkin, A. (1980) *Science* **210**, 1081.

Ashkin, A. & Dziedzic, J. M. (1981) *Appl. Opt.* **20**, 1803.

Ashkin, A., Dziedzic, J. M. & Stolen, R. H. (1981) *Appl. Opt.* **20**, 2299.

Baer, T. (1987) *Opt. Lett.* **12**, 392.

Baker, B. B. & Copson, E. T. (1950) *The Mathematical Theory of Huygens' Principle*, 2nd Edn.. Oxford University Press.

Barber, P. W. & Chang, R. K., eds. (1988) *Optical Effects Associated with Small Particles*. Singapore: World Scientific.

Barrette, J. & Alamanos, N. (1985) *Nucl. Phys.* A**441**, 733.

Barton, J. P., Alexander, D. R. & Schaub, S. A. (1988) *J. Appl. Phys.* **64**, 1632.

Barton, J. P., Alexander, D. R. & Schaub, S. A. (1989) *J. Appl. Phys.* **65**, 2900.

Barut, A. O. & Calogero, F. (1962) *Phys. Rev.* **128**, 1383.

Baseia, B. & Nussenzveig, H. M. (1984) *Optica Acta* **31**, 39.

Bath, M. (1979) *Introduction to Seismology*, 2nd Edn.. Basel: Birkhäuser Verlag.

Beck, D. (1962) *J. Chem. Phys.* **37**, 2884.

Ben-Menahem, A. & Singh, S. J. (1981) *Seismic Waves and Sources*. New York: Springer-Verlag.

Benincasa, D. S., Barber, P. W., Zhang, J.Z., Hsieh, W. F. & Chang, R. K. (1987) *Appl. Opt.* **26**, 1348.

Bergé, P., Pomeau, Y. & Vidal, C. (1987) *Order within Chaos: Towards a Deterministic Approach to Turbulence*. New York: Wiley.

Bernstein, R. B. (1966) in *Advances in Chemical Physics*, ed. J. Ross, vol. 10, p. 313. New York: Wiley.

Berry, M. V. (1966) *Proc. Phys. Soc.* **89**, 479.

Berry, M. V. (1969) *J. Phys.* **B2**, 381.

Berry, M. V. & Mount, K. E. (1972) *Rep. Prog. Phys.* **35**, 315.

Berry, M. V. & Upstill, C. (1980) in *Progress in Optics*, ed. E. Wolf, vol. 18, 275. Amsterdam: North-Holland.

Blatt, J. M. & Weisskopff, V. (1952) *Theoretical Nuclear Physics*. New York: Wiley.

Bohren, C. F. & Huffman, D. R. (1983) *Absorption and Scattering of Light by Small Particles*. NewYork: Wiley.

Bollini, C. J. & Giambiagi, J. J. (1962) *Nuovo Cimento* **26**, 619.

Bollini, C. J. & Giambiagi, J. J. (1963) *Nuovo Cimento* **28**, 341.

Born, M. & Wolf, E. (1959) *Principles of Optics*. New York: Pergamon.

Bosanac, S. (1978a) *Molec. Phys.* **35**, 1057.

Bosanac, S. (1978b) *Molec. Phys.* **36**, 453.

Bosanac, S. (1978c) *J. Math. Phys.* **19**, 789.

Bosanac, S. (1979) *Phys. Rev. A* **19**, 125.

Bouwkamp, C. J. (1954) *Rep. Prog. Phys.* **17**, 35.

Boyer, C. B. (1987) *The Rainbow: from Myth to Mathematics*. Princeton University Press.

Brandt, J. C. (1968) *Publ. Astron. Soc. Pacific* **80**, 25.

Bremmer, H. (1949) *Terrestrial Radio Waves*. Amsterdam: Elsevier.

Bricard, J. (1940) *Ann. Phys. (Paris)* **14**, 148.

Brill, D. & Gaunard, G. (1987) *J. Acoust. Soc. Am.* **81**, 1.

Brink, D. M. (1985) *Semi-classical Methods for Nucleus–Nucleus Scattering*. Cambridge University Press.

Brink, D. M. & Takigawa, N. (1977) *Nucl. Phys.* **A279**, 159.

Bryant, H. C. & Cox, A. J. 1966) *J. Opt. Soc. Am.* **56**, 1529.

Bryant, H. C. & Jarmie, N. (1968) *Ann. Phys. (NY)* **47**, 127.

Bryant, H. C. & Jarmie, N. (1974) *Sci. Am.* **231**, 60.

Bucerius, H. (1946) *Optik* **1**, 188.

Buck, U. & Pauly, H. (1971) *J. Chem. Phys.* **54**, 1929.

Buck, U. (1974) *Rev. Mod. Phys.* **46**, 369.

Buck, U. (1975) *Adv. Phys. Chem.* **30**, 313.

Buck, U. (1988) in *Atomic and Molecular Beam Methods*, ed. G. Scoles, vol. 1, p. 499. Oxford University Press.

Campillo, A. J., Eversole, J. D. & Lin, H.-B. (1991) *Phys. Rev. Lett.* **67**, 437.

Cerveny, V. & Ravindra, R. (1971) *Theory of Seismic Head Waves*. University of Toronto Press.

Chandrasekhar, S. (1950) *Radiative Transfer*. Oxford: Clarendon Press.

Chapman, C. H. & Phinney, R. A. (1972) in *Methods in Computational Physics*, ed. B. A. Bolt, vol. 12, p. 165. New York: Academic.

Chen, G., Acker, W. P., Chang, R. K. & Hill, S. C. (1991) *Opt. Lett.* **16**, 117.

Chester, C., Friedman, B. & Ursell, F. (1957) *Proc. Camb. Phil. Soc.* **53**, 599.

Ching, S. C., Leung, P. T. & Young, K. (1990) *Phys. Rev.* **A41**, 5026.

Chitanvis, S. M. & Cantrell, C. D. (1989) *J. Opt. Soc. Am.* **B6**, 1326.

Chylek, P. (1976) *J. Opt. Soc. Am.* **66**, 285.

Chylek, P., Ramaswamy, V., Ashkin, A. & Dziedzic, J. M. (1983) *Appl. Opt.* **22**, 2302.

Chylek, P., Pendleton, J. D. & Pinnick, R. G. (1985) *Appl. Opt.* **24**, 3940.

Chylek, P. (1990) *J. Opt. Soc. Am.* **A7**, 1609.

Collins, P. D. B. (1977) *An Introduction to Regge Theory and High-Energy Physics*. Cambridge University Press.

Connor, J. N. L. & Marcus, R. A. (1971) *J.Chem. Phys.* **55**, 5636.

Connor, J. N. L., Delos, J. B. & Carlson, C. E. (1976) *Molec. Phys.* **31**, 1181.

Connor, J. N. L. (1980) in *Semiclassical Methods in Molecular Scattering and Spectroscopy*, ed. M. S. Child, p. 45. Dordrecht: Reidel.

Connor, J. N. L., Farrelly, D. & Mackay, D. C. (1981) *J. Chem. Phys.* **74**, 3278.

Conwell, P. R., Barber, P. W. & Rushforth, C. K. (1984) *J. Opt. Soc. Am.* **A1**, 62.

Cormier, V. F. & Richards, P. G. (1977) *J. Geophys.* **43**, 3.

Crifo, J. F. (1988) in *Optical Particle Sizing*, eds. G. Gouesbet and G. Gréhan, p.529. New York: Plenum Press.

da Silveira, R. (1973) *Phys. Lett.* **45B**, 211.

da Silveira, R. (1990) in Neskovic (1990), p. 103; cf. also Orsay Report IPNO/TH88-69.

Dave, J. V. (1969) *Appl. Opt.* **8**, 155.

Davis, G. E. (1955) *J. Opt. Soc. Am.* **45**, 572.

de Alfaro, V. & Regge, T. (1965) *Potential Scattering*. Amsterdam: North-Holland.

de Bruijn, N. G. (1958) *Asymptotic Methods in Analysis*. Amsterdam: North-Holland.

Debye, P. J. (1908) *Physik. Z.* **9**, 775.

Debye, P. J. (1909) *Ann. Phys.* **30**, 57.

Delbar, T., Grégoire, G., Paic, G., Ceuleneer, R., Michel, F., Vanderpoorten, R., Budzanowski, A., Dabrowski, H., Freindl, L., Grotowski, K., Micek, S., Planeta, R., Strzalkowski, A. & Eberhard, K. A. (1978) *Phys. Rev.* **C18**, 1237.

Descartes, R. (1637) *Les Météores*, appendix to *Discours de la Méthode*. Leyden: Jean Maire.

Dingle, R. B. (1973) *Asymptotic Expansions: their Derivation and Interpretation*. New York: Academic.

Di Salvo, E. & Viano, G. A. (1977) *Nuovo Cimento* **A42**, 49.

Di Salvo, E. & Viano, G. A. (1982) *Nuovo Cimento* **A71**, 261.

Di Salvo, E. (1983) *Nuovo Cimento* **A74**, 427.

Doornbos, D. J. & Mondt, T. C. (1979a) *Geophys. J. Roy. astr. Soc.* **57**, 353.

Doornbos, D. J. & Mondt, T. C. (1979b) *Geophys. J. Roy. astr. Soc.* **57**, 381.

Duwalo, G. & Jacobs, J. A. (1959) *Can. J. Phys.* **37**, 109.

Engdahl, E. R. (1968) *Science* **161**, 263.

Esposito, L. W., Knollenberg, R. G., Marov, M. Ya., Toon, O. B. & Turco, R. P. (1983) in *Venus*, eds. D. M. Hunten, L. Colin, T. M. Donahue and V. I. Moroz, p. 484. University of Arizona Press.

Eversole, J. D., Lin, H.-B., Huston, A. L. & Campillo, A. J. (1990) *J. Opt. Soc. Am.* **A7**, 2159 (1990).

Fahlen, T. S. & Bryant, H. C. (1968) *J. Opt. Soc. Am.* **58**, 304.

Feynman, R. P., Leighton, R. B. & Sands, M. (1964) *The Feynman Lectures in Physics*, 3 vols. Reading: Addison-Wesley.

Feynman, R. P. & Hibbs, A. R. (1965) *Quantum Mechanics and Path Integrals*. New York: McGraw-Hill.

Fiedler-Ferrari, N. & Nussenzveig, H. M. (1981) in *Proceedings of the II Brazilian Meeting on Particles and Fields*, p. 73. São Paulo: Brazilian Physical Society.

Fiedler-Ferrari, N. (1983) Ph. D. thesis, University of São Paulo.

Fiedler-Ferrari, N. & Nussenzveig, H. M. (1987) *Part. Charact.* **4**, 147.

Fiedler-Ferrari, N., Nussenzveig, H. M. & Wiscombe, W. J. (1991) *Phys. Rev.* **A43**, 1005.

Flax, L., Dragonette, L. R. & Uberall, H. (1978) *J. Acoust. Soc. Am.* **63**, 723.

Flax, L., Gaunard, G. G. & Uberall, H. (1981) in *Physical Acoustics*, eds. W. P. Mason and R. N. Thurston, vol 15, p. 191. New York: Academic.

Fock, V. A. (1948) *Phil. Mag.* **39**, 149.

Fock, V. A. (1965) *Electromagnetic Diffraction and Propagation Problems*. Oxford: Pergamon.

Ford, K. W. & Wheeler, J. A. (1959) *Ann. Phys. (NY)* **7**, 259.

Frahn, W. E. (1985) *Diffractive Processes in Nuclear Physics*. Oxford University Press.

Fraser, A. B. (1983) *J. Opt. Soc. Am.* **73**, 1626.

Franz, W. (1957) *Theorie der Beugung Elektromagnetischer Wellen*. Berlin: Springer.

Friedlander, F. G. (1958) *Sound Pulses*. Cambridge University Press.

Friedman, F. L. & Weisskopf, V. F. (1955) in *Niels Bohr and the Development of Physics*, eds. W. Pauli *et al.*, p. 134. London: Pergamon Press.

Fuller, R. C. (1975) *Phys. Rev.* **C12**, 1561.

Fuller, R. C. & Moffa, P. J. (1977) *Phys. Rev.* **C15**, 266.

Garrett, C. G. B., Kaiser, W. & Bond, W. L. (1961) *Phys. Rev.* **124**, 1807.

Gerard, A. (1979) *Int. J. Engng. Sci.* **17**, 313.

Gerard, A. (1980) *Int. J. Engng. Sci.* **18**, 583.

Goldstein, H. (1957) *Classical Mechanics*. Reading: Addison-Wesley.

Goody, R. M. & Yung, Y. L. (1989) *Atmospheric Radiation*. Oxford University Press.

Goos, F. & Hänchen, H. (1947) *Ann. Phys. Lpz.* (6) **1**, 333.

Gouesbet , G. & Gréhan, G., eds. (1988) *Optical Particle Sizing: Theory and Practice.* New York: Plenum Press.

Greenler, R. (1980) *Rainbows, Halos and Glories.* Cambridge University Press.

Guimarães, L. G. & Nussenzveig, H. M. (1992) to appear in *Optics Comm.*

Gustafsson, T. (1983) in *Atomic Physics 8* , eds. I. Lindgren, A. Rosén and S. Svanberg, p. 355. New York: Plenum Press.

Hansen, J. E. & Arking, A. (1971) *Science* **171**, 669.

Hansen, J. E. & Hovenier, J. W. (1974) *J. Atmos. Sci.* **31**, 1137.

Hansen, J. E. & Travis, L. D. (1974) *Space Sci. Rev.* **16**, 527.

Hanyga, A. , ed. (1985) *Seismic Wave Propagation in the Earth.* Amsterdam: Elsevier.

Haroche, S. (1984) in *New Trends in Atomic Physics*, eds. G. Grynberg and R. Stora, p.193. Amsterdam: Elsevier.

Hayter, A. (1973): *A Voyage in Vain: Coleridge's Journey to Malta in 1804.* London: Faber & Faber.

Heisenberg, W. (1971) *Physics and Beyond.* New York: Harper and Row.

Henyey, L. G. & Greenstein, J. L. (1941) *Astrophys. J.* **93**, 76.

Hill, S. C., Rushforth, C. K., Benner, R. E. & Conwell, P. R. (1985) *Appl. Opt.* **24**, 2380.

Hill, S. C. & Benner, R. E. (1986) *J. Opt. Soc. Am. B***3**, 1509.

Hsieh, W.-F., Zheng, J.-B. & Chang, R. (1988) *Opt. Lett.* **13**, 497.

Humphreys, W. J. (1964) *Physics of the Air.* New York: Dover.

Hussein, M. S., Nussenzveig, H. M., Villari, A. C. C.& Cardoso, J. L. (1982) *Phys. Lett. B***114**, 1.

Hussein, M. S. & McVoy, K. W. (1984) in *Progress in Nuclear and Particle Physics*, ed. D. Wilkinson, vol. 12, p. 103. Oxford: Pergamon Press.

Ince, E. L. (1956) *Ordinary Differential Equations.* New York: Dover.

Irvine, W. M. (1965) *J. Opt. Soc. Am.* **55**, 16.

Jackson, J. D. (1975) *Classical Electrodynamics*, 2nd Edn. New York: Wiley.

Jones, D. S. (1964) *The Theory of Electromagnetism*, p. 516. Oxford: Pergamon Press.

Juan, J. & Ulloa, A. (1748) *Relación Histórica del Viaje a la América Meridional.* Madrid: A. Marin.

Kamran, M. (1984) *Phys. Rep.* **108**, 275.

Kawabata, K., Coffeen, D. L., Hansen, J. E., Lane, W. A., Sato, M. & Travis, L. D. (1980) *J. Geophys. Res.* **85**, 8129.

Keller, J. B. (1958) in *Calculus of Variations and its Applications, Proc. Symp. Appl. Math.,* vol. 8, ed. L. M. Graves, p. 27. New York: McGraw-Hill.

Keller, J. B. (1962) *J. Opt. Soc. Am.* **52**, 116.

Keller, J. B., Lewis, R. M. & Seckler, B. D. (1956) *Commun. Pure Appl. Math.* **9**, 207.

Kerker, M. (1969) *The Scattering of Light and Other Electromagnetic Radiation.* New York : Academic.

Khare, V. & Nussenzveig, H. M. (1974) *Phys. Rev. Lett.* **33**, 976.

Khare, V. (1975) Ph. D. thesis, University of Rochester (available from University Microfilms, Inc.).

Khare, V. & Nussenzveig, H. M. (1977a), *Phys. Rev. Lett.* **38**, 1279.

Khare, V. & Nussenzveig, H. M. (1977b) in *Statistical Mechanics and Statistical Methods in Theory and Application,* ed. U. Landman, p. 723. New York: Plenum.

Khare, V. (1982) in *Electromagnetic Surface Modes,* ed. A. D. Boardman, p. 417. New York: Wiley.

Kleyn, A. W. (1987) *Comments At. Mol. Phys.* **19**, 133.

Knollenberg, R. G. & Hunten, D. M. (1980) *J. Geophys. Res.* **85**, 8039.

Kong, P., Mason, E. A. & Munn, R. J. (1970) *Am. J. Phys.* **38**, 294.

Können, G. P. (1985) *Polarized Light in Nature.* Cambridge University Press.

Korsch, H. J. & Thylwe, K. E. (1983) *J. Phys.* **B16**, 793.

Krasnopolsky, V. A. (1985) *Planet. Space Sci.* **33**, 109.

Kurizki, G. & Nitzan, A. (1988) *Phys. Rev.* **A38**, 267.

Lai, H. M., Leung, P. T. & Young, K. (1988) *Phys. Rev.* **A37**, 1597.

Lai, H. M., Leung, P. T., Young, K., Barber, P. W. & Hill, S. C. (1990) *Phys. Rev.* **A41**, 5187.

Lai, H. M., Leung, P. T. & Young, K. (1990) *Phys. Rev.* **A41**, 5199.

Langer, R. E. (1937) *Phys. Rev.* **51**,669.

Langley, D. S. & Marston, P. L. (1984) *Appl. Opt.* **23**, 1044.

Lee, S. Y. (1978) in *Nuclear Physics with Heavy Ions and Mesons,* eds. R. Balian, M. Rho and G. Ripka, vol. 1, p. 55. Amsterdam: North-Holland.

Leontovich, M. A. (1944) *Bull. Acad. Sci. USSR* **8**, 16.

Levinson, N. (1949) *Kgl. Danske Videnskab. Selskab., Mat.–fys. Medd.* **25**, no. 9.

Levy, B. R. & Keller, J. B. (1959) *Comm. Pure Appl. Math.* **12**, 159.

Levy, B. R. & Keller, J. B. (1960) *Canad. J. Phys.* **38**, 128.

Lin, H.-B., Huston, A. L., Justus, B. L. & Campillo, A. J. (1986) *Opt. Lett.* **11**, 614.

Linke, F. & Möller, F. (1961) *Handbuch der Geophysik,* vol. 8. Berlin: Gebrüder Bornträger.

Lipperheide, R. (1987) *Nucl. Phys.* **A469**, 190.

Lisle, I. G., Parlange, J. Y., Rand, R. H., Hogarth, W. L., Braddock, R. D. & Gottlieb, H. P. (1985) *Phys. Rev. Lett.* **55**, 555.

Lock, J. A. (1987) *Appl. Opt.* **26**, 5291.

Lock, J. A. (1988) *J. Opt. Soc. Am.* **A5**, 2032.

Lock, J. A. & Woodruff, J. R. (1989) *Appl. Opt.* **28**, 523.

Logan, N. A. (1965) *Proc. IEEE* **53**, 773.

Lorenz, L. (1890) *Kgl. Danske Vidensk. Selsk. Skrifter* **6**, 1.

Lötsch, H. K. V. (1971) *Optik* **32**, 116, 189, 299, 553.

Ludwig, D. (1969) *Comm. Pure Appl. Math.* **22**, 715.

Lynch, D. K. & Futterman, S. N. (1991) *Appl. Opt.* **30**, 3538.

Lyubimov, V. A. (1978) *Sov. Phys.–Usp.* **20**, 691.

McCartney, E. J. (1976) *Optics of the Atmosphere.* New York: Wiley.

McDonald, J. E. (1963) *Am. J. Phys.* **31**, 282.

McVoy, K. W., Khalil, H. M., Shalaby, M. M. & Satchler, G. R. (1986) *Nucl. Phys.* A**455**, 118.

McVoy, K. (1990) in Neskovic (1990), p. 75.

Malyuzhinets, G. D. (1959) *Sov. Phys.–Usp.* **69**, 749.

Marov, M. Ya., Lystsev, V. E., Lebedev, V. N., Lukashevich, N. L. & Shari, V. P. (1980) *Icarus* **44**, 608.

Marov, M. Ya., Byshev, B. V., Baranov, B. P., *et al.* (1983) *Cosmic Res.* **21**, 269.

Marston, P. L. (1979) *J. Opt. Soc. Am.* **69**, 1205 [*ibid.* **70**, 353 (E)].

Marston, P. L. & Kingsbury, D. L. (1981) *J. Opt. Soc. Am.* **71**, 192, 917 (E).

Marston, P. L., Johnson, J. L., Love, S. P. & Brim. B. L. (1983a), *J. Opt. Soc. Am.* **73**, 1658.

Marston, P. L., Williams, K. L. & Hanson, T. J. B. (1983b) *J. Acoust. Soc. Am.* **74**, 605.

Marston, P. L. & Trinh, E. H. (1984) *Nature* **312**, 529.

Marston, P. L. (1985) *Opt. Lett.* **10**, 588.

Marston, P. L. (1991) *Appl. Opt.* **30**, 3479, 3549.

Marty, C. (1983) *Z. Phys.* A**309**, 261.

Mason, E. A., Nyeland, C., van der Biesen, J. J. H. & van den Meijdenberg, C. J. N. (1982) *Physica* **116A**, 133.

Maze, G., Taconet, B. & Ripoche, J. (1981) *Phys. Lett.* **84A**, 1981.

Maze, G. & Ripoche, J. (1983) *J. Acoust. Soc. Am.* **73**, 41.

Melrose, R. & Taylor, M. E. (1986) *Comm. Partial Differential Equations* **11**, 599.

Messiah, A. (1959) *Mécanique Quantique.* Paris: Dunod.

Mie, G. (1908) *Ann. d. Phys.* **25**, 377.

Minnaert, M. (1954) *The Nature of Light and Color in the Open Air.* New York: Dover.

Mott, N. F. & Massey, H. S. W. (1965) *The Theory of Atomic Collisions*, 3rd Edn., Oxford University Press.

Mount, K. E. (1973) *J. Phys.* B**6**, 1397.

Mullen, J. M. & Thomas, B. S. (1973) *J. Chem. Phys.* **58**, 5216.

Myers, W. D.(1974) in *Proc. Int. Conf. on Reactions between Complex Nuclei*, eds. R. L. Robinson *et al*, vol. 2, p. 1. Amsterdam: North-Holland.

Neskovic, N., ed. (1990) *Rainbows and Catastrophes* . Belgrade: Boris Kidric Institute of Nuclear Science.

Neubauer, W. G. (1973) in *Physical Acoustics*, vol. 10, eds. W. P. Mason and R. N. Thurston, p. 61. New York: Academic.

Newton, I. (1704) *Opticks.* London: Royal Society.

Nobel Foundation (1965) *Nobel Lectures in Physics*. Amsterdam: Elsevier.

Nörenberg, W. & Weidenmüller, H. A. (1976) *Introduction to the Theory of Heavy-Ion Collisions*. Berlin: Springer.

Nussenzveig, H. M. (1959) *An. Acad. Brasil. Cienc.* **31**, 515.

Nussenzveig, H. M. (1965) *Ann. Phys. (NY)* **34**, 23.

Nussenzveig, H. M. (1969a) *J. Math. Phys.* **10**, 82.

Nussenzveig, H. M. (1969b) *J. Math. Phys.* **10**, 125.

Nussenzveig, H. M. (1972a) *Causality and Dispersion Relations*. New York: Academic.

Nussenzveig, H. M. (1972b) *Phys. Rev.* **D6**, 1534.

Nussenzveig, H. M. (1973) *Introduction to Quantum Optics*. New York: Gordon and Breach.

Nussenzveig, H. M. (1977) *Sci. Am.* **236**, 116.

Nussenzveig, H. M. (1979) *J. Opt. Soc. Am.* **69**, 1068.

Nussenzveig, H. M. & Wiscombe, W. J. (1980a) *Phys. Rev. Lett.* **45**, 1490.

Nussenzveig, H. M. & Wiscombe, W. J. (1980b) *Opt. Lett.* **5**, 455.

Nussenzveig, H. M. & Wiscombe, W. J. (1987) *Phys. Rev. Lett.* **59**, 1667.

Nussenzveig, H. M. (1988) *J. Phys.* **A21**, 81.

Nussenzveig, H. M. (1989a) in *Atomic Physics 11*, eds. S. Haroche, J. C. Gay and G. Grynberg, p. 421. Singapore: World Scientific.

Nussenzveig, H. M. (1989b) *Comments At. Mol. Phys.* **23**, 175.

Nussenzveig, H. M. (1990) in *Coherence and Quantum Optics VI*, eds. J. H. Eberly, L. Mandel and E. Wolf, p. 821. New York: Plenum Press.

Nussenzveig, H. M. & Wiscombe, W. J. (1991) *Phys. Rev.* **A43**, 2093.

Nye, J. F. (1984) *Nature* **312**, 531.

Olver, F. W. J. (1974) *Asymptotics and Special Functions*. New York: Academic.

Orear, J. (1964) *Phys. Rev. Lett.* **12**, 112.

Ostrowski, A., Tiereth, W. Brandl, D., Basrak, Z. & Voit, H. (1989) *Phys. Lett.* **B232**, 46.

Patashinskii, A. Z., Pokrovskii, V. L. & Khalatnikov, I. M. (1963) *Sov. Phys.–JETP* **17**, 1387.

Pato, M. P. & Hussein, M. S. (1990) *Phys. Rep.* **189**, 127.

Pearcey, T. (1947) *Phil. Mag.* **37**, 311.

Pernter, J. M. & Exner, F. M. (1910) *Meteorologische Optik*. Vienna: Braumüller.

Petiau, G. (1955) *La Théorie des Fonctions de Bessel*. Paris: CNRS.

Pinnick, R. G., Biswas, A., Chylek, P., Armstrong, R. L., Latifi, H., Creegan, E., Srivastava, V., Jarzembski, M. & Fernández, G. (1988) *Opt. Lett.* **13**, 494.

Poincaré, H. (1910) *Rend. Circ. Mat. Palermo* **29**, 169.

Poston, T. & Stewart, I. N. (1978) *Catastrophe Theory and its Applications*. London: Pitman.

Probert-Jones, J. R. (1984) *J. Opt. Soc. Am.* **A1**, 822.

Pulfrich, C. (1888) *Ann. Phys. Chem. (Leipzig)* **33**, 209.

Put, L. W. & Paans, A. M. J. (1977) *Nucl. Phys.* A**291**, 93.

Qian, S.-X. & Chang, R. K. (1986) *Phys. Rev. Lett.* **56**, 926.

Qian, S.-X., Snow, J. B., Tzeng, H.-M. & Chang, R. K. (1986) *Science* **231**, 486.

Ragent, B. & Blamont, J. (1980) *J. Geophys. Res.* **85**, 8089.

Ray, B. (1923) *Nature* **111**, 183.

Rayleigh, Lord (1877) *The Theory of Sound*, vol. 1, p. 70. London: Macmillan & Co. [reprinted, New York: Dover (1945)].

Rayleigh, Lord (1899) *Phil. Mag.* **47**, 345.

Richards, P. G. (1973) *Geophys. J. R. astr. Soc.* **35**, 243.

Robin, L. (1958) *Fonctions Sphériques de Legendre et Fonctions Sphéroïdales*, vol. 2. Paris: Gauthier-Villars.

Rubinowicz, A. (1917) *Ann. d. Phys.* **53**, 257.

Rubinowicz, A. (1965) in *Progress in Optics*, ed. E. Wolf, vol. IV, p. 201. Amsterdam: North-Holland.

Saunders, M. J. (1970) *J. Opt. Soc. Am.* **60**, 1359.

Schiff, L. I. (1968) *Quantum Mechanics*, 3rd Edn. New York: McGraw-Hill.

Schöbe, W. (1954) *Acta Math.* **92**, 265.

Scholte, J. G. J. (1956) *Kon. Ned. Met. Inst. Publ.* **65**, 9.

Schrempp, B. & Schrempp, F. (1977a) *Nuovo Cimento Lett.* **20**, 95.

Schrempp, B. & Schrempp, F. (1977b) *Phys. Lett.* B**70**, 88.

Schrempp, B. & Schrempp, F. (1980) *Nucl. Phys.* B**163**, 397.

Schrödinger, E. W. (1928) *Collected Papers on Wave Mechanics.* London: Blackie & Son.

Schutte, A., Bassi, D., Tommasini, F. & Scoles, G. (1972) *Phys. Rev. Lett.* **29**, 979.

Schwartz, L. (1966) *Mathematics for the Physical Sciences.* Reading: Addison-Wesley.

Shipley, S. T. & Weinman, J. A. (1978) *J. Opt. Soc. Am.* **68**, 130.

Simpson, H. J. & Marston, P. L. (1991) *Appl. Opt.* **30**, 3468, 3547.

Snow, J. B., Qian, S.-X. & Chang, R. K. (1985) *Opt. Lett.* **10**, 37.

Sommerfeld, A. (1954) *Optics.* New York: Academic.

Streifer, W. & Kodis, R. D. (1964) *Quart. Appl. Math.* **21**, 285.

Takigawa, N. & Lee, Y. S. (1977) *Nucl. Phys.* A**292**, 173.

Taylor, J. R. (1974) *Nuovo Cimento* B**23**, 313.

Teng, T. L. & Richards, P. G. (1969) *J. Geophys. Res.* **74**, 1537.

Thom, R. (1972) *Stabilité Structurelle et Morphogénèse.* Reading: Benjamin.

Thylwe, K. E. (1983) *J. Phys.* A**16**, 1141.

Thylwe, K. E., & Connor, J. N. L. (1985) *J. Phys.* A**18**, 2957.

Titchmarsh, E. C. (1937) *Introduction to the Theory of Fourier Integrals*, 2nd Edn. Oxford University Press.

Toennies, J. P., Welz, W. & Wolf, G. (1979) *J. Chem. Phys.* **71**, 614.

Tzeng, H.-M., Wall. K. F., Long, M. B. & Chang, R. K. (1984) *Opt. Lett.* **9**, 499.

Tricker, R. A. R. (1970) *Introduction to Meteorological Optics.* New York: Elsevier.

Uberall, H. (1973) in *Physical Acoustics*, eds. W. P. Mason and R. N. Thurston, vol. 10, p. 1. New York: Academic.

Ursell, F. (1965) *Proc. Camb. Phil. Soc.* **61**, 113.

van den Biesen, J. J. H., Hermans, R. M. & van den Meijdenberg, C. J. N. (1982) *Physica* **115A**, 396.

van den Biesen, J. J. H.(1988) in *Atomic and Molecular Beam Methods*, ed. G. Scoles, vol. 1, p. 472. Oxford University Press.

van de Hulst, H. C. (1946-8) *Recherches Astron. Obs. d'Utrecht* **11** (Utrecht thesis).

van de Hulst, H. C. (1947) *J. Opt. Soc. Am.* **37**, 16.

van de Hulst, H. C. (1957) *Light Scattering by Small Particles.* New York: Wiley.

van de Hulst, H. C. (1980) *Multiple Light Scattering*, vols. 1 and 2. New York: Academic.

van de Hulst, H. C. & Wang, R. T. (1991) *Appl. Opt.* **30**, 4756.

Volz, F. (1961) *Der Regenbogen*, in *Handbuch der Geophysik*, vol. 8, eds. F. Linke and F. Möller. Berlin: Gebrüder Bornträger.

Walker, J. (1976) *Am. J. Phys.* **44**, 421.

Walker, J. (1977) *Sci. Am.* **237**, 138.

Walker, J. (1978a) *Sci. Am.* **239** (4), 146.

Walker, J. (1978b) *Sci. Am.* **239** (4), 179.

Wang, R. T. & van de Hulst, H. C. (1991) *Appl. Opt.* **30**, 106.

Watson, G. N. (1918) *Proc. Roy. Soc. (London)* **A95**, 83.

White, F. P. (1922) *Proc. Roy. Soc. (London)* **A100**, 505.

Whittaker, E. T. (1927) *Analytical Dynamics.* Cambridge University Press.

Wigner, E. P. (1955) *Phys. Rev.* **98**, 145.

Williams, E. J. (1945) *Rev. Mod. Phys.* **17**, 217.

Williams, K. L. & Marston, P. L. (1985a) *J. Acoust. Soc. Am.* **78**, 722.

Williams, K. L. & Marston, P. L. (1985b) *J. Acoust. Soc. Am.* **78**, 1093.

Wiscombe, W. J. (1979) *Mie scattering calculations: advances in technique and fast, vector-speed computer codes*, NCAR/TN-140. Boulder: National Center for Atmospheric Research.

Wiscombe, W. J. (1980) *Appl. Opt.* **19**, 1505.

Wu, T. T. (1956) *Phys. Rev.* **104**, 1201.

Zhang, J.-Z., Leach, D. H. & Chang, R. K. (1988) *Opt. Lett.* **13**, 270.

Zhang, J.-Z. & Chang, R. K. (1989) *J. Opt. Soc. Am.* **B6**, 151.

# Index